Canal Lifts and Inclines of the World

Hans-Joachim Uhlemann

translated and edited by

Mike Clarke

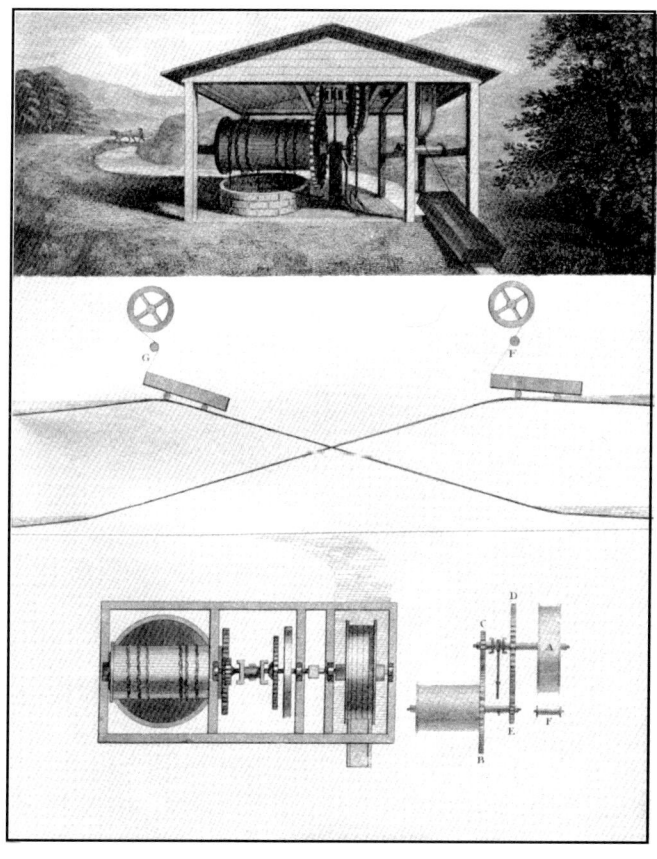

INTERNAT

2002

Published by Internat, 2002
Registered Office: Crabtree Hall, Mill Lane, Lower Beeding, Horsham
RH13 6PX

ISBN 0-9543181-0-2 (hb)
ISBN 0-9543181-1-0 (sb)

British Library Cataloguing in Publication Data.
A catalogue record of this book is available from the British Library.

Design and layout by Mike Clarke, Accrington
Printed by Cromwell Press, Trowbridge

A Note about Dimensions

British/Metric

Imperial measurements have been quoted where they were used originally, but with a metric conversion. The 'ton' is used throughout as it is so similar to the metric 'tonne'.

Weight
 1 kilogram = 2.2 lbs
 1 lb = 0.37 kilograms

Length
 1 metre = 39.37 inches
 1 km = 0.62 miles
 1 inch = 2.54 cm
 1 foot = 0.31 metres
 1 mile = 1.61 km

Pressure
 1 atmosphere/bar = 14.7 lbs/in^2

Prussia

Translations from German articles often use the Prussian Fuß which has been translated as a British foot.

Weight
 1 Centner = 110.23 lbs

Length
 1 Zoll = 1.03 inches
 1 Fuß = 12.36 inches
 1 Ruthe = 12.36 feet

CONTENTS

Foreword

This book about canal lifts and inclines, possibly the most fascinating and impressive structures on inland waterways, is the result of ten years of research. This took in numerous sources, particularly those from Germany and England, including the extensive technical literature which was scoured for references.

I was particularly interested in the opinions of outstanding early experts, but also valued more recent quotations, and I would like to thank those involved for their co-operation. This includes my many friends and colleagues from the waterway administration who have supported my work. In particular Dr.-Ing. Martin Eckoldt, the founder of the Studienkreis für Geschichte des Wasserbaus, der Wasserwirtschaft und der Hydrologie, for his valuable references and allowing access to his personal documents.

Thanks also to the Gemeentearchief in Amsterdam, the Westfalian Industry Museum, Dortmund, the Science Museum Library in London, the Mitchell Library in Glasgow, and the Shropshire Records and Research Unit,

Shrewsbury, for providing valuable material. Cordial thanks also go to Mr. John Good from the Canadian Heritage Parks for his photos and supplementary information on the Canadian lifts, and to Mr. William J. Moss of the Canal Society of New Jersey for the reproductions of historic photographs of the inclined planes on the Morris Canals. My English friends Ron Oakley and Edward Paget-Tomlinson for their photos; Dr.-Ing. Hans Rohde and Prof. Dr. Otfried Wagenbreth for permission to copy from their publications, as well as Dr. Ing. Jaroslav Kubec for his interesting information and photos.

My friend Mike Clarke, whose unselfish help and valuable advice, particularly concerning the English lifts, were invaluable and the idea and preparatory work for the chronology of lifts and inclines since 1777, I thank especially.

But not least is the understanding and help of my wife and daughter, and the generous support of the publishers, without which this work would not have been completed.

Hans-Joachim Uhlemann
Author
Die Geschichte der Schiffshebewerke
DSV-Verlag
ISBN 3-88412-291-6

Foreword to the English Edition

I am sure that so many people walking along the canals take for granted the fact that boats of all sizes can pass through the countryside. I suppose the only group of people who do not take it for granted are the boaters themselves. The system of 'locking' boats up, over and down hills is well known, but possibly what is not so well known are those extraordinary pieces of machinery that actually lift boats up and over inclines where the building of locks is just not possible.

This book is truly a fountain of information about lifts from 1700 BC to the present day! It is packed with information, diagrams, maps and photographs to illuminate what could be seen as a dry subject. I am fascinated by the thought that boats have been navigated inland for nearly 3,000 years, and it would have been impossible for them to do so without the invention of so many diverse boat lifts.

It is interesting to read how the expertise has been built up over the years. The expert knowledge of engineers, surveyors, boaters and all geographers combine to make a unique and specialised service industry. This industry has contributed so greatly to not only the enjoyment of so many people today, but indeed to making the movements of goods on inland waterways not only a hope or remote possibility, but a reality, thereby contributing to business and industry throughout the world.

David Suchet

Acknowledgements

As with all books of this type, many people have been involved with its preparation. Besides those mentioned by the author in his original Foreword, I would particularly like to thank Kate Bonson, Bernard Weiss, David Edwards-May and John Boyes for their help in proof-reading the text.

It would be impossible to produce a book such as this without assistance. Friends not just in Britain but across Europe have helped with my research into European inland waterways, including Andris Biedrins (Latvia), Vadim Mikheyev (Russia), Prof. Stanslaw Januszewski, Robert Kola, Jan Gustaw Jurkiewicz and Artur Zbiegieni (Poland), Jur Kingma (The Netherlands), Jaques de la Garde (France), and Prof. Muskatirovic (Yugoslavia), as well as those already mentioned, particularly in Germany, by Hans-Joachim Uhlemann.

Mike Clarke

Fig. 1 The Fossa Carolina, probably the first waterway to be built after the Roman period.

Fig. 2 The Alster Navigation, near Hamburg, still has flash locks. The small boats and canoes which use the waterway today are carried round the locks by means of slipways like the one on the right. They are, in effect, simple inclined planes.

Fig. 3 The Moulin-Brulé staircase locks on the Canal de Briare were built early in the seventeenth century. This was only the second canal which had locks with ground paddles, sluices which were built into the lock walls. It was an important development as it allowed locks to be built deeper. Previously, sluices had been built into the lock gates, and if the rise was too great, water passing through them would fall onto the boat in the lock. Lifts were another way to get around the problem of safely overcoming a large rise.

1 THE ORIGIN OF WATERWAYS

The first transport routes, apart from the sea, were streams and rivers, so the earliest human settlements were often found alongside them. As the 'road' was available, it needed only the 'vehicles' to transport people and goods.

The first inland waterways were unregulated watercourses whose geography governed boat size. For centuries, unimproved streams and rivers were used for transport. Early inland vessels were smaller than those of today, so they could go into areas no longer accessible by large modern inland waterway craft.

After the Roman period, the first European canal was the Fossa Carolina, designed and built by Charlemagne in 793AD to link the head waters of the rivers Danube and Main and thus forming a water route from the Black Sea to the North Sea. It was a simple channel across the watershed, part of which can still be seen. However, there does not seem to have been any deepening of the rivers associated with it.

To improve the navigation of small rivers, the earliest man-made structures were flash locks. They were movable barriers which blocked the river completely or partially and could be opened to provide sufficient water for shipping.

When a barrier was opened, boats could float over shallow downstream river sections, which were otherwise unnavigable, on the wave of water released. Because the boats travelled at almost the same speed as the water, it was difficult to steer them, whilst those sailing upstream had to be hauled laboriously against the current. In 1398, the rivers Delvenau and Stecknitz in Germany were made navigable by flash locks, creating the famous Stecknitz Canal between the Elbe at Lauenburg and the Trave at Lübeck. Around this time, flash locks were also used to make the Alster navigable from Hamburg.

The origin of fully man-made canal systems only became possible with the introduction of the chamber or pound lock, an invention which appeared in China c.1000AD and later, almost simultaneously, in The Low Countries, Italy, France and Germany.

The first description of one is attributed to the Italian, Leone Battista Alberti, in his work *'De re aedificatoria'* (Book 10, Cap. 12), which he presented to Pope Nicholas V in the year 1452, as reported in the Eusebian Chronicle.

Make the barriers double, in that you cut the

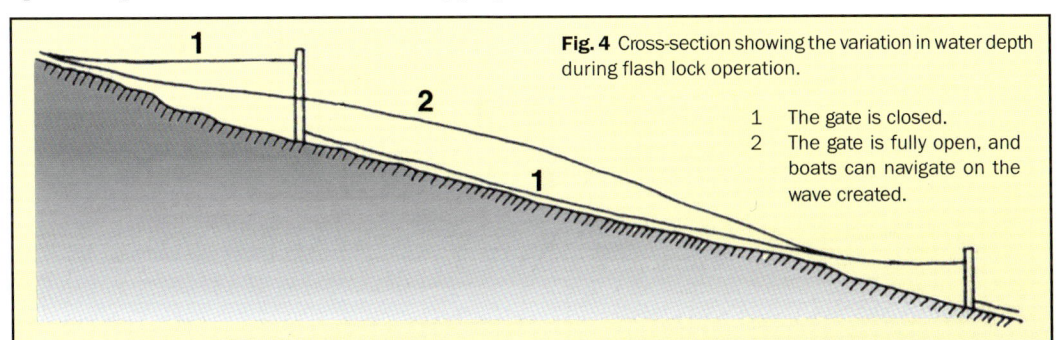

Fig. 4 Cross-section showing the variation in water depth during flash lock operation.

1   The gate is closed.
2   The gate is fully open, and boats can navigate on the wave created.

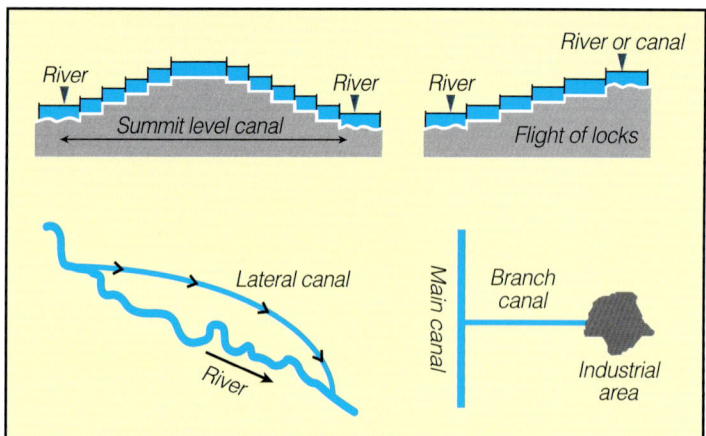

Fig. 5 Cross-section of a summit level canal and a flight of locks, with plans of a lateral and a branch canal.

river at two places and leave a space which holds the length of a boat, so that, when a boat sailing upstream arrives here, the lower barrier must be closed and the upper opened; if it is however a boat travelling downstream, in that case the upper is closed and the lower opened. The boat is allowed in this way, if the water is let off, to continue downstream. The water left will be held back by the upper barrier.

Using chamber locks it was now possible to build a canal across hills. From the seventeenth century, summit-level canals were built across the high ground between river-basins, early examples being the French Canal de Briare, opened in 1642, and the Canal du Midi of 1681, and the Prussian Friedrich Wilhelms Canal or 'Neue Graben', constructed between 1662 and 1668 to link the Elbe to the Oder.

Summit-level canals are the 'crown' of canal building. Other types include lateral canals, built parallel to an unnavigable river, branch canals linking a source of raw materials or an industrial area to another waterway, and those connecting two canals at different levels.

Rivers and streams whose navigation could not be improved by traditional river regulation works, such as groins and riverbed excavation, could now be canalised. The water level was controlled by means of weirs with locks so that constant water depth could be maintained for boats. The locks were built either in the weir itself or on separately constructed lock cuts.

There are thousands of locks across the world, but there are only a few dozen inclines and lifts. Where are they, and how were they built? This book gives some of the answers.

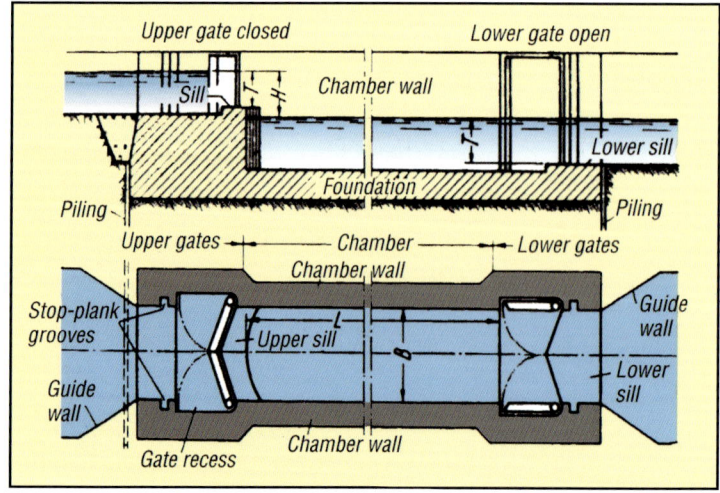

Fig. 6 A simple chamber lock. T: Depth over the sill, H: Rise, B: Breadth, L: Length.

2   THE ORIGINS OF LIFT CONSTRUCTION

It may seem incredible, but inclined planes are the oldest means of assisting boats to pass watersheds and variations in land level. They have been used by mankind since the onset of water transport and they carried the then relatively small and light vessels from one water level to the next just by manpower.

Schleswig-Holstein

It is often said that the Vikings carried their long boats overland during invasions and could thus reach lakes not directly linked to navigable waters. It is also suspected that they crossed the Schleswig-Holstein peninsula near the ancient merchant city of Haithabu (it was destroyed in 1066) with small dug-out boats. The rivers Eider and Treene may have been navigable up to Hollingstedt and from there they could have reached Haithabu, which lay on the Schlei, by carrying their boats along an eight kilometre long track. It is more probable that they took their small boats up the Rheider Au towards the military road (Heerweg), taking them out of the water near to Klein Rheide, carrying them overland on rollers along the four kilometre Kograben and then relaunching them into the Selker Noor.[1] We will never know for certain, but it is highly probable when one looks at other parts of the world where, very much earlier, the transport of boats had taken place over such watersheds.

Since manhandling boats must have been hard work and only suitable for the smallest of vessels, people soon began to look for other ways to transport boats from one waterway to another more effectively. For this the inclined plane seemed particularly effective. Together

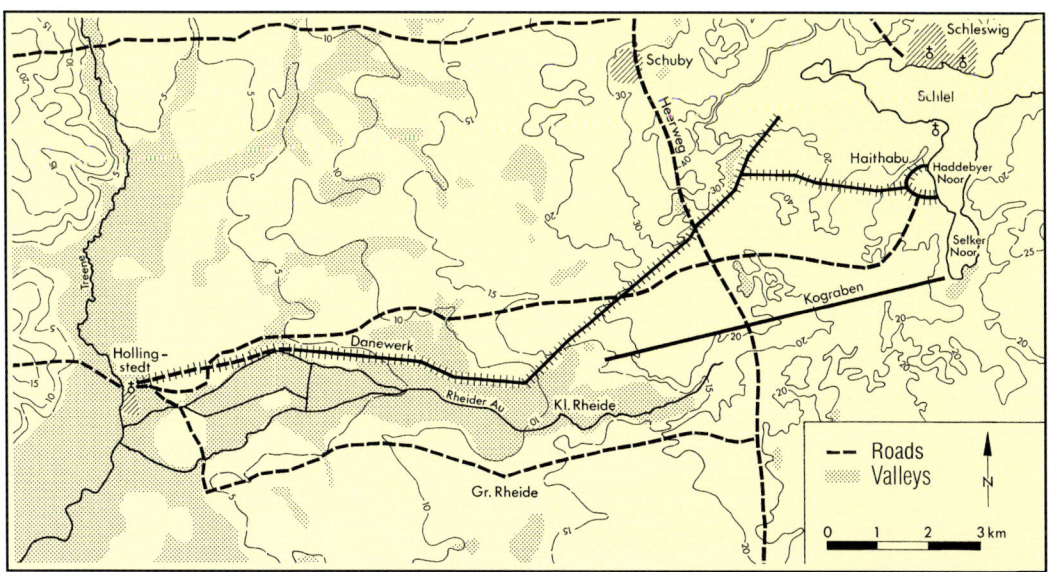

Fig.7 Possible route for boats, dating from the 9th to 11th century AD, between the North Sea and the Baltic lying across the Schleswig-Holstein peninsula.

with levers, wedges and rollers, it is a basic part of ancient mechanical technology.[2]

The Nile

In Egypt, a three or four kilometre long slipway may have been constructed around 1700 BC for avoiding the second Nile cataract near to Wadi Halfa.[3] Due to rapids, the Nile was only passable here with ease during the season of floods, between the end July and the end November. Therefore, between Mirgissa, 20 km south of Wadi Halfa, and Abusir Rock, they built a gently-sloping embankment made from Nile mud. It had a recessed slipway in which were fixed transverse pieces of wood one Egyptian yard (about 51 cm) apart. Nile mud is very slippery when damp, so a man was sent ahead of each boat to pour water on the slipway. Experiments have shown that it is difficult to stop the boats once in motion.

Passing the rapids at Mirgissa was of great importance as stone for the Pharaohs' pyramids and temples was transported on the Nile from upper to lower Egypt. Unfortunately, the place where possibly the first inclined plane in the world was constructed is today covered by the waters of the Aswan Dam.[4]

The Diolkos of Corinth

It has also been suggested by Strabon in his fourth book that, around 600 BC, Periandros, the tyrant of Corinth, built Diolkos (meaning 'to slide through something') over the 6.4 km broad isthmus of Corinth. This stone-lined way has tracks with deep grooves, gauge about one and a half metres, which suggests that boats were carried on wagons. These were probably about five metres long and three and a half metres broad.[5] Although several ancient texts suggest that cargo ships were transported over the isthmus, it is more likely that it was only used by warships. Cargo ships were probably unloaded and their goods carried over the isthmus and then reloaded into other ships. It is known that Octavius, later the first Roman Emperor Augustus (63 BC-14 AD), used the inclined plane to carry his fleet of 250 ships over the isthmus after the battle of Aktium in 31 BC, in which he destroyed the naval power of Egypt. There was a similar feat in the year 1453, when

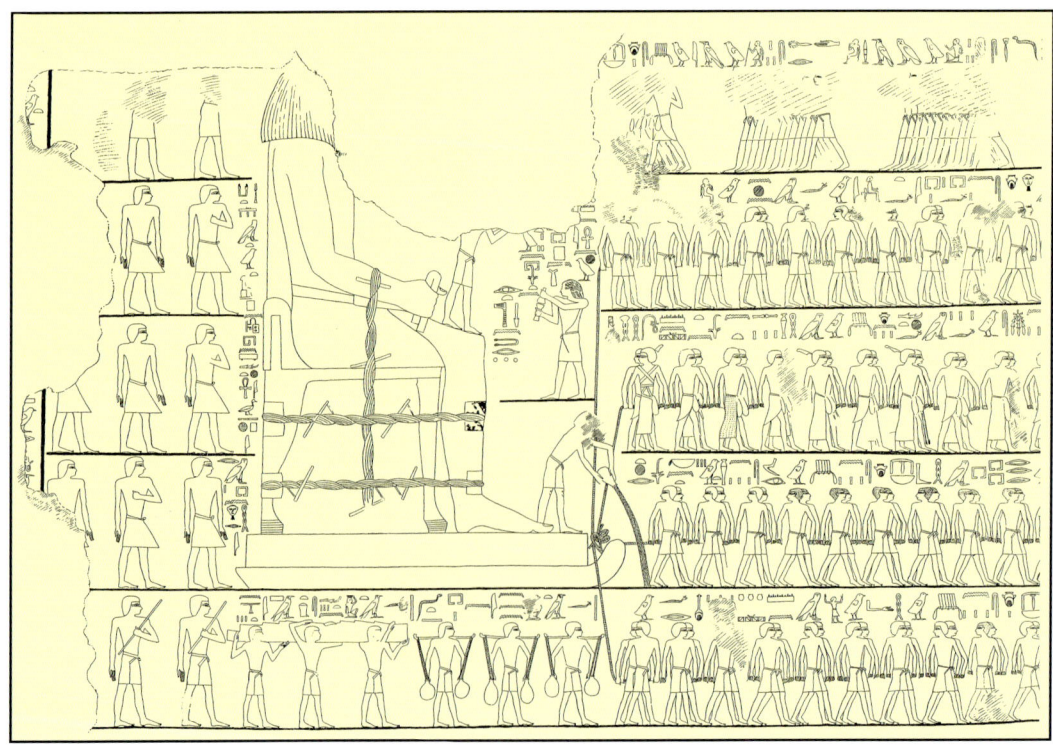

Fig. 8 Moving the Colossus from Bersheh. In the middle of the picture, a figure is wetting the slipway.

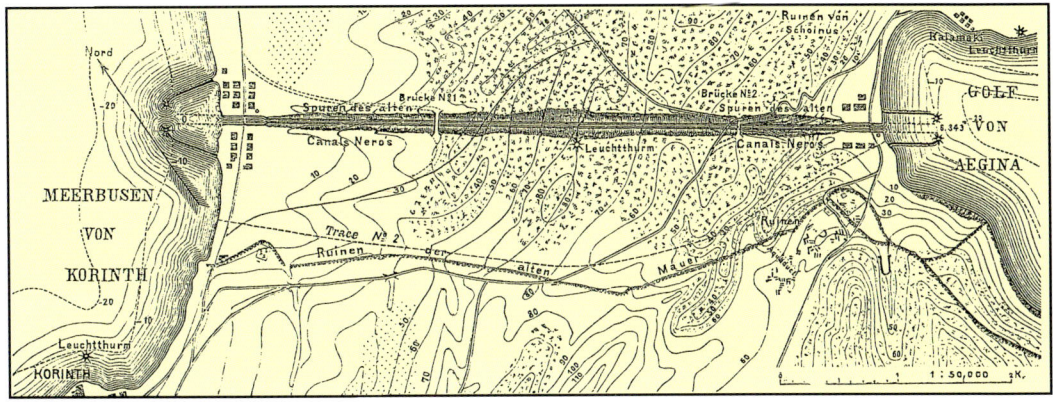

Fig. 9 A map of the Isthmus of Corinth, across which ships passed by means of the Diolkos for around 1800 years. Nero sought to build a canal across this isthmus, but the idea was not completed until 1882-93 with the construction of the Corinth Canal. The Diolkos began to the south of the western end of the canal, crossed its route and then continued eastwards parallel to it.

the Turks crossed the headland of the Golden Horn to begin the conquest of Byzantium, as recorded in the poetry of Stefan Zweig.[6] It was, however, an unusual use of inclines for the time.

The Diolkos saved ships from a long and dangerous voyage around the southern tip of the Peloponnus and provided the city of Corinth with a large income from transit duties. They continued to be used until the twelfth century.[7] A section was found in 1956 and has been preserved,[8] though much of the route was probably destroyed by the Corinth Canal which opened in 1893.

China

On early Chinese canals, simple inclines were often used. Small boats were pulled by a capstan up a solid clay incline or a wooden slide to overcome a change in level.[9]

Hadfield[10] mentions that such boat lifts were used in China in 348AD and probably earlier, with boats being moved by means of oxen-powered capstans. Before 385AD, there were seven such inclines on the Ssu River section of the Grand Canal (see map on page 132), which used either oxen or large groups of men.[11] Many foreign travellers tell of such inclines. In 1073, a Japanese monk named Jójin found, close to Hang-chow, an incline which operated with the help of a capstan, and in 1307 Rashin al-Din wrote about the Grand Canal: *'Boats of such a size are also pulled up by a kind of machine and lowered back into the water on the other side*[12]*.'*

At the end of the eighteenth century, English travellers had visited inclines on which the

Fig. 10 The excavated remains of the Diolkos. In the background is the south-western end of the Corinth Canal

11

Fig. 11 A simple Chinese incline.

Fig. 12 Cross-section of a simple incline.

Loam

boats were hauled by two capstans opposite each other across the upper end of the incline. The capstans were held upright between four stone columns, while their axles were supported on a lower stone between two of the stone columns. On each capstan, four large bars were manned by between 12 and 16 persons, and the time for raising a boat was between two and a half and three minutes.[13] The height difference of the individual inclines was between 1.8 and 3 metres and the angle of slope was from 45 to 50 degrees.[14]

The 'Overtooms' of the Low Countries

A similar development took place in the Netherlands. Numerous rivers flow here, and people have had to defend themselves from the earliest times against the devastating storm tides of the North Sea. As a result, the Dutch

Fig. 13 A contemporary drawing of a Chinese incline.

have become outstanding hydraulic engineers. The old motto of the coastal inhabitants: *"Wer nicht will weichen, der muss deichen"* (Those who do not want to be soaked, they must leave), is nowhere more applicable than in this low country. Coastal and flood protection is closely connected with the development of locks. To be able to let off water which had accumulated behind a dike into the sea, it was necessary to provide gaps in the dike. In these gates were fitted which were opened during the ebb tide and closed during the flood. Thus emerged the duikersluizen. Passing boats through these locks was, however, a laborious task since it was possible only if the water levels were the same within and without. The loss of time which resulted from having to wait to pass through the lock soon forced a search for better alternatives. One was the chamber (or pound) lock, a second the construction of inclines similar to those in China on which boats were pulled over the dam or dike. These inclines were called 'overtoomen', 'overhaalen' or 'overdrachten'.

The first overtooms were built in the twelfth century in the Low Countries close to Ypres,[15] and they could be used by boats of eight tons capacity.[16] The Nieuwedamme overtoom on the IJzer at Ypres was built in 1167 and consisted of two inclines with, at its summit, a treadmill-powered capstan. A chain was wound over this and then connected to the boats which were to be raised or lowered.[17]

The oldest overtoom in the Netherlands was probably that built in Spaarndam around 1200[18] and the oldest illustration is of the one at Hoorn, which appears on a map of 1530.[19]

The exact number of overtooms constructed in the Netherlands is unknown but there were certainly a considerable number. With their simple but archaic technology, they served primarily for the transport of small vessels which were loaded with agricultural products. An exception was the overtoom at Zaandam which was in use between 1609 and 1718, and over which seagoing vessels up to 7.32 metres (24 feet) in width were hauled. So that they could slide more easily on the wooden slipways,

Fig. 14 A simple Dutch incline or overtoom.

these were smeared with fat or grease. Even then, 24 to 30 people were necessary to turn the incline's three capstans.

Many overtooms survived until the beginning of the twentieth century; there were even two electrified ones in Amsterdam dating from 1916 and 1923, the latter only being removed in 1965. All that is left today is an old lullaby[20] and two which are preserved.

The Venhuizen overtoom

To find this overtoom, take the main road from Enkhuizen to Hoorn and, near Stede Broec, turn off in the direction of Venhuizen. The overtoom crosses the road just before it reaches the village. There are two horizontal wooden capstans on either side of a road at the top of wood-lined inclines up which small boats were formerly pulled over wooden rollers. The waterways at either end of the overtoom are no longer used by boats. The wooden capstans and the inclines have been reconstructed and to help visitors to understand their operation a small boat lies in front of one of the capstans.

The 'Blaue Molen' overtoom

The remains of another overtoom can be found in South Holland at Rijpwetering, which lies half way between Amsterdam and Leiden on the A4/E 10. This overtoom is right beside the *Blauwe Molen* on the Blauwe polder of Rijpwetering. Although the main road is only about 200 metres away, the mill and overtoom form a typical rural Dutch landscape. The overtoom itself comprises a wooden capstan and one short and one long grass slipway. There is a small boat, just as at Venhuizen.

Broekerhaven

There used to be an overtoom in the West Friesland village of Broekerhaven but today it has been replaced by a boat lift, now preserved as a technical monument.

The harbour at Broekerhaven was laid out in 1415 on the edge of the Zuiderzee in order to make the villages of Grootebroek, Lutjebroek and Bovenkarspel independent of Enkhuizen.

Fig. 15 The overtoom between the Kostverlorenvaart and the Slotervaart in Amsterdam in 1820.

Fig. 16 Reconstructed overtoom at Venhuizen, south-west of Enkhuizen, in West Friesland.

The overtoom was built in the seventeenth century to connect the polder canals to the harbour. Small boats were pulled over it on a bed of wet clay and after passing the summit they slipped back into the water by gravity. The difference in level between the polder of Het Grootslag and the Zuiderzee was between two and three metres. As with several overtooms, a steam-engine was installed around 1900 to work the winches. After the First World War the incline was replaced by a vertical lift without a caisson. (see page 88)

Fig. 17 & 18 Two views of the remains of the overtoom at Blauwe Molen, near the A4/E10 motorway half-way between Leiden and Amsterdam.

The Dutch only built two small vertical lifts. The flat lands and the increasing size of boats meant locks were better overcoming variations in water level. The Netherlands is an important area for water technology and developed, in the overtooms, interesting inclined lifts.

Russian 'Woloki'

In other countries primitive ways of hauling boats over a summit have also been used for centuries. For example, inclines existed in the European part of Russia, marked as 'woloki' (from 'wolotschit': to drag) on maps, and small boats sailing from the Volga to rivers flowing into the Baltic Sea were hauled over them. The name of the town of Vyschni Volotschek is a reminder of the existence of such inclines.[21]

Another example may exist on the Solovki Islands where, in the sixteenth century, monks constructed canal systems linking many lakes in the islands. They may also have built a simple incline. In the same area Peter the Great hauled naval ships southwards from the White Sea along the line of today's Baltic-White Sea Canal allowing them to engage the Swedish fleet in the Baltic more quickly than if they had sailed around the north of Norway.

The 'Wagon of Zafosina'

The best described and most spectacular incline of the late Middle Ages was the *Wagon of Zafosina* which Vittorio Zonca (1568-1602), the city engineer of Padua, described in his work *Novo Teatro di Machine et Edificii* of 1607.

Fusina or Zafosina is at the end of the Canale di Brenta. Leupold describes this incline as follows:[22]

"At Lizzafusina, five Venetian miles from Venice, where fresh water meets sea water, boats cannot pass because of a strong wooden dam which is built there. This prevents the canal, which lies several feet higher, from emptying into the sea. Over this dam, boats sitting on a large sled with rollers and powered by horses turning a capstan are carried from the canal to the sea and from the sea to the canal. This machine is called by local people the 'Wagon of Lizzafusina'.

This wagon is made from sawn timber, of which two pieces, N O, are long, and have at each end iron rings, P Q, on to which the hooks of the ropes are attached. The other two pieces of wood are shorter, which together equal the length of the other pieces. In the middle of them are two more pieces of wood, R S, just as long as the short ones, and they are fitted well into each other. Into this four-cornered construction are four small wheels with strong iron ferrules which have a diameter the width of a shoe, and their thickness three-quarters. The rest of this machine is on the ground; so that the wagon or the sled can run into and out of the water. Nearby is a capstan, constructed with a drive formed from two bars erected cross-wise to which a horse can be attached. They drive a gearwheel, on whose axle the rope to the iron

Fig.19 One of the canals in the Solovki Islands.

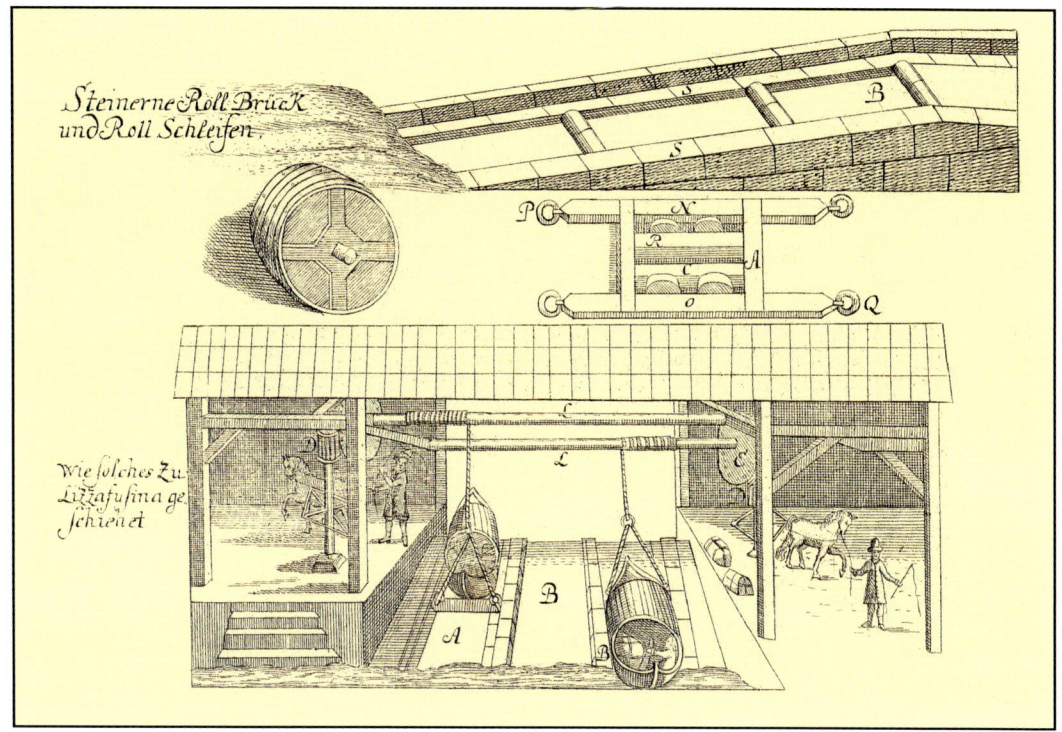

Fig. 20 The 'Wagon of Zafosina' (Lizzafusina) by Vittorio Zonca.

hooks, which haul the boat, is wrapped around.

Note that the horse going round on the right-hand side hauls the boats from the sea, while the other horse, on the left-hand side, hauls them out of the river or canal. Between the river Brenta and the sea, where the wagon operates, a wall is built, and a roof over it: there are also two slightly raised stone guides for the wheels of the wagon which are somewhat broader than the wheels. Alongside the guides on both sides leading into the water, are pavements made from exceedingly large and hard stones, such that the guides are not spoiled if driven over by the wagon. Other individual features can be understood from the figure."

German designs

Prussia began to develop fully its inland navigations in the mid-eighteenth century and engineers, such as Sturm, Leupold and Fäsch, discussed the use of Dutch overtooms for carrying boats between waterways in their books published early in the century. Since boats were often pulled over rollers, they called these structures *'Rollbrücken'* (rolling bridges).

Although the works of these authors provided instructive illustrations, their ideas were not put into practice.

Sturm's observations, in which he speculates on the gradual disappearance of *'Rollbrücken'* in Holland, set down the main advantages, in his opinion for their use in Germany, which are still worthy of consideration today. He made a comparison as to whether it is better to use a lock or an incline and came to the following conclusion:

"I have to maintain accordingly after very carefully considering everything and have no shyness in presenting this sentence to the public: if an important person has small rivers which power many mills and he wants to make them navigable it will be better if he to this end allows inclines to be built rather than if he constructs locks."[23]

The main grounds for this conclusion were:

1 That by the use of inclines no water supply is needed, so that the entire water flow of the river benefits the mills. If one thinks about the 'weir' war between millers and

17

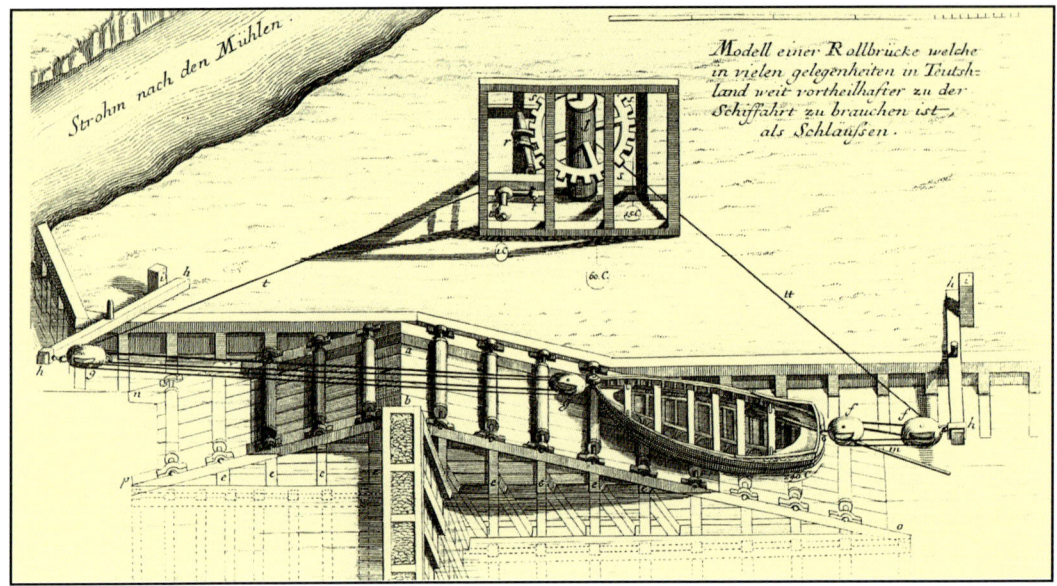

Fig. 21 Drawing of a design for an incline by Sturm.

lock keepers of about 200 years ago on Friedrich Wilhelms Canal[24] and elsewhere, such as on the Thames in England, then Sturm's suggestions are worthy of some consideration.

2   A comparison of building costs comes out clearly in favour of inclines, cost being one of the most important considerations taken into account throughout the whole history of navigational construction works.

However, in the same way as early English canal builders, Sturm did not realise that inland water transport would create a dramatic rise in industry and the effects this would have. In England, the small narrow canals built to keep costs down could not be enlarged to meet the growing demands for transport in the mid-nineteenth century. With regard to inclines and lifts, these could only carry small boats until the development of new materials and power sources of the late nineteenth century, and locks were probably the best solution to moving boats between different canal levels in the eighteenth and early nineteenth century, except for isolated cases and canals.

3 LIFTS OF THE INDUSTRIAL REVOLUTION

After enduring a long period of stagnation, the eighteenth century saw a new chapter in the history of inclined and vertical lifts. Until the middle of the nineteenth century this was almost exclusively confined to England, then the foremost industrial country. It is interesting to note that the beginning of the industrial revolution around 1760 was closely associated with the development of their inland waterway system. The construction of new inclines began in 1777 on Ducart's Canal in Northern Ireland. Within a few decades, England, which had previously played an insignificant role in the development of waterway engineering, became one of the leaders in this area.

Lock, Inclined Lift or Vertical Lift?

Inclined and vertical lifts are certainly the most impressive structures on modern waterways. *"In contrast to locks, which are static structures with hydraulic processes familiar to waterway engineers, lifts are dynamic structures using various different hydraulic, mechanical and electric techniques."*[25]

Even though they are extremely complicated, they are still used in particular circumstances, though locks are now replacing some lifts in Germany. Although there are many thousands of locks across the world, there are only about two dozen lifts. They are used when either only a small water supply is available (the water usage of a lift is minimal) or if large variations in level must be overcome in a short distance. In 1984 Partenscky[26] suggested that for level changes of 25 metres or over lifts should be considered and above 35 metres they are better

than chamber locks. However, recent studies of lock design have questioned these findings.

When speed of operation is important, the advantage lies with lifts. An average speed of 12.6 m/min is obtained on the vertical lift at Scharnebeck (near Lüneburg) which entered service in 1975, as compared with 3 m/min for modern chamber locks. The question as to whether lifts or locks are to be used can only be clarified by economic investigation into their construction and maintenance costs. Besides these criteria, the geological foundations are also an important consideration.

Types of Lift

In general, for both inclined and vertical lifts, there are two possible methods for raising or lowering a boat; those where it is transported out of the water - dry lifts; and those where the boat floats in a caisson filled with water - wet lifts.

The dry type dominated the development of simple lifts. On the earliest ones the boat was hauled over a slide, on mid-period ones it was supported on rollers, and on the final designs it rested on a platform or in a cradle fitted with wheels. The dry system was mainly used on inclines, and in only a few exceptional cases were boats lifted vertically out of the water. On wet lifts, where the boat floats in a caisson, this either moved vertically or along an inclined plane. More recently, inclined lifts have been constructed where boats float in a wedge of water moved up or down an inclined channel.

The dry lift system, which developed over several centuries, is rarely used today in Europe, although it does have some advantages. For both

Fig. 22 A typical English tub boat. Such boats were often used on early dry lifts to restrict the weight being moved.

energy requirement for moving a caisson is very low. This balance can be provided either by counterweights or by a second caisson operating in the opposite direction. The latter is a *twin* lift, while a *double* lift is where the caissons operate independently but alongside each other.

Lifts are best categorised by their angle of operation. They are either inclined or vertical lifts, the former also known as inclined planes.

Inclined lifts can be subdivided, depending on whether the boat is transported parallel to, or crosswise to, its usual direction of travel, that is longitudinal or transverse operation. The water-slope is a special form of longitudinal operation.

Vertical lifts can be divided as follows:

1 Flotation lifts, where the weight of caisson and boat is supported by air-tight tanks immersed in water-filled wells.

2 Water or oil hydraulic lifts where the weight of the caisson is taken by pistons operated by hydraulic pressure.

3 Counterweight lifts where separate weights counterbalance the boat or caisson.

4 Special forms of lift using compressed air, balance beams or drums which remained theoretical until the Falkirk Wheel entered service in Scotland.

5 Diving locks where the boat floats in an enclosed container which can be submerged in a water-filled shaft.

types of lift the ratio between payload (the boat and its cargo) and total load (including the caisson, if used, and the carriage) is better when the boat is removed from the water; just 15-35% when using a water-filled caisson, more than 60% when the boat is out of the water. Because of this, on dry lifts the slipways or lift framework can be of smaller dimensions and thus cheaper. The crucial factor in the use of caissons today is that the safety of the boat being transported is improved and, above all, that the time for the operation of the lift is less.

With a water-filled caisson the weight to be moved remains almost constant, since the boat displaces water equal to its gross weight (boat plus load). For this reason it is easy to provide an almost perfect weight balance. This is not possible when the boat is removed from the water. By ensuring an exact weight balance the

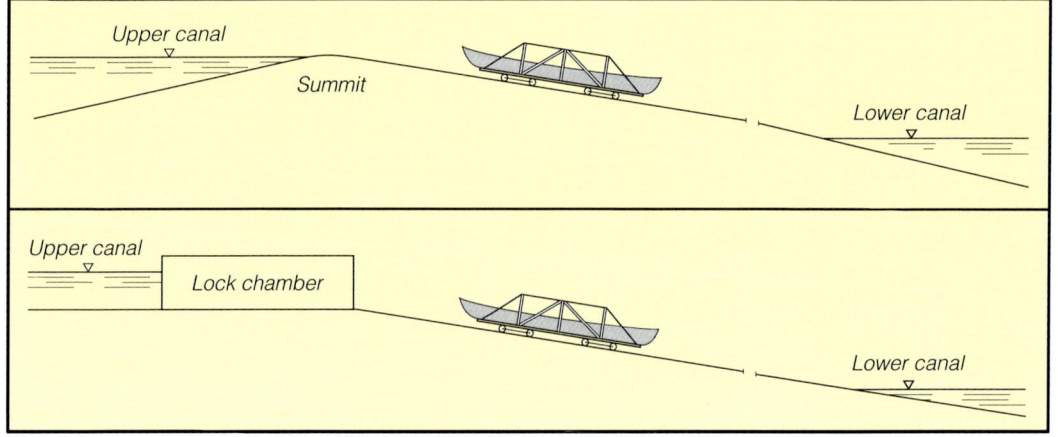

Fig. 23 Inclined planes without a caisson: above with a 'dry' summit level, below with a chamber lock at the upper end.

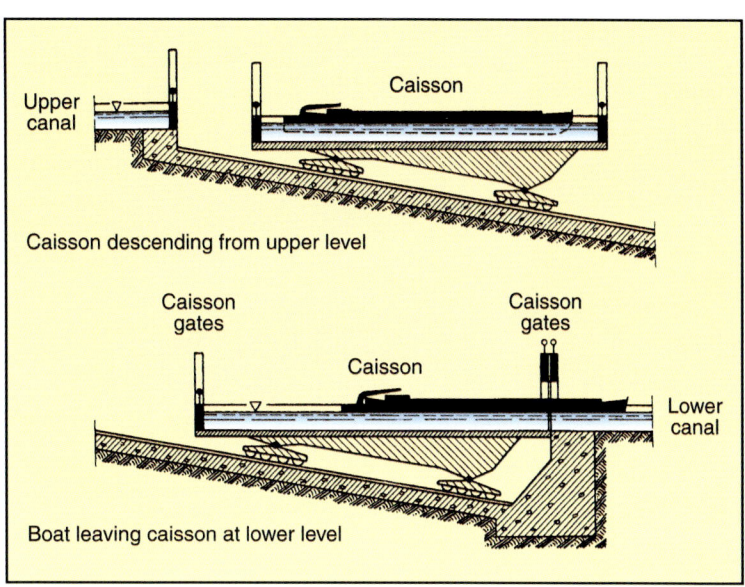

Fig 24 An inclined plane with a water-filled caisson.

INCLINED PLANES

The best slope for a longitudinal inclined plane is between 1:15 and 1:50. As it becomes steeper, the height of the outer end of the caisson and its carriage has to be increased to keep the boat level, making construction both difficult and expensive, and in these situations transverse operation or a vertical lift must be used. This is particularly important for modern boats of 110 metres length; the small size of boats used up to the end of the nineteenth century required small carriages which could be built without difficulty for steep inclines.

Dry longitudinal inclines
Ducart's Canal

Lift construction at the start of the industrial revolution began with longitudinal inclines, on which the boats were carried dry.[27] The first lifts of this kind were built on Ducart's Canal in Northern Ireland. It allowed coal to be carried to Dublin from mines at Drumglass, via the River Blackwater, Lough Neagh and the Newry Canal, to Newry from where the final part of its journey was over the Irish Sea.

Ducart, who designed and built the canal, came from the Sardinian Kingdom where he had been an engineer, so he could have had knowledge of Italian waterway engineering, such as the *Wagon of Zafosina*. His original plan, which was approved in 1767 by the Irish Parliament, proposed building a four mile long canal on two levels to the mines. Coal was to be carried in twelve tub-boats, each carrying

Fig 25 The remains of a bridge over one of the Coalisland Canal inclines which were called 'dry hurries'.

21

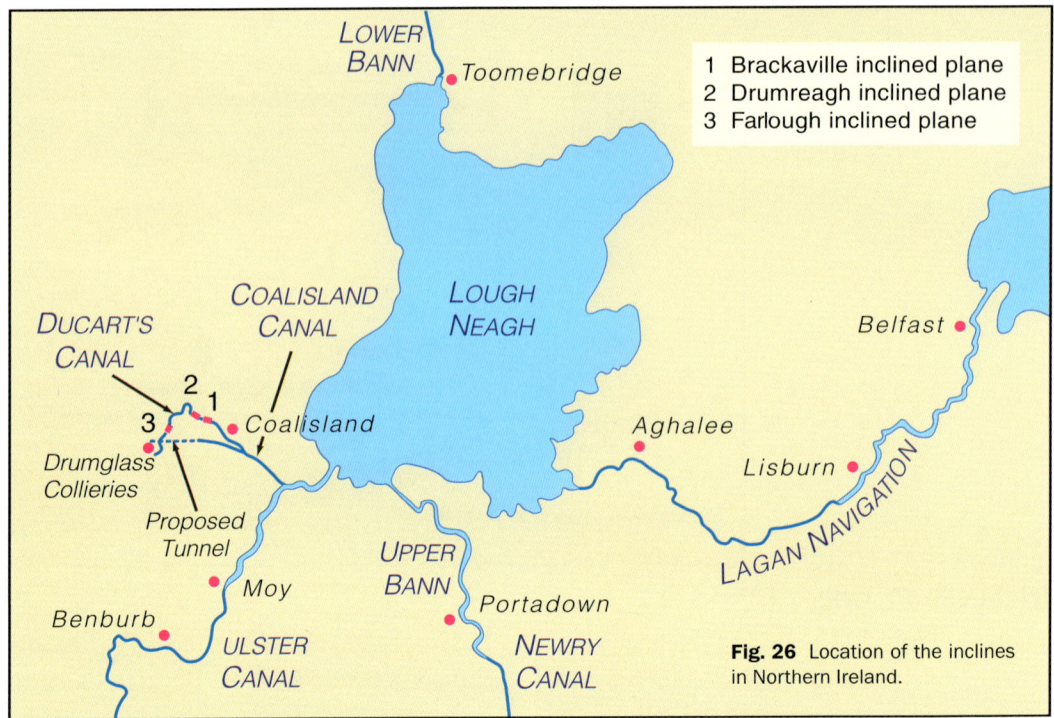

Fig. 26 Location of the inclines in Northern Ireland.

Within the map:

LOWER BANN

Toombridge

1 Brackaville inclined plane
2 Drumreagh inclined plane
3 Farlough inclined plane

COALISLAND CANAL

LOUGH NEAGH

Belfast

DUCART'S CANAL

2 1
3
Coalisland

Aghalee

Lisburn

Drumglass Collieries

LAGAN NAVIGATION

Proposed Tunnel

UPPER BANN

Moy

Portadown

Benburb

ULSTER CANAL

NEWRY CANAL

one ton (Tub-boats usually only worked on small canals). The difference of 148 feet (45 metres) between the two canal levels was to be surmounted by a shaft, down which boxes of coal could be lowered. In November 1767, Ducart changed his plans and decided to construct three inclined planes, at Brackaville, Drumreagh and Farlough, with rises of 55, 65 and 70 feet respectively (16.76, 19.8 and 21.34 metres). Similar to but larger than Dutch overtooms, they were designed with wooden ramps and rollers. A water wheel powered the winding mechanism which raised the boats. In 1773, work on the inclines was interrupted after an inspection of the canal by the well-known engineers, William Jessop and John Smeaton. The latter thought that the scheme was imprudent and proposed instead that the inclined planes should be doubled, a loaded boat going down drawing up an empty one. He also recommended that the capacity of the boats should be increased from one to two tons and that a capstan be installed at the upper end of the incline which would pull the boats over the summit. The boats were 10 feet

(3.05 metres) long and 4 feet 6 inches (1.37 metres) broad, with one end square and at the other pointed. In practice, the boats did not slide smoothly over the rollers, and Ducart fitted rails and a four-wheel carriage on which the boats rested. Horses were used to power the capstan.

Ducart's Canal was ready by 1777, though it could not be used until the Coalisland Canal had been built. Ducart died a year later and may not have realised his inclines were a failure. Probably they were too steep and too advanced for the technology of the time. By 1787 they had been dismantled and coal was taken to the Coalisland Canal by road. However, masonry and a bridge over an incline still survives.

Ketley incline

Eleven years after the completion of Ducart's Canal, an inclined plane was constructed at Ketley, near Ironbridge, which can probably be called the first successful lift of the modern period.

Early English and Irish inclines, and the first German lifts described later, were built not for general cargo, but for traffics related to specific industries. Ketley incline was no exception, being set out …*on a small isolated canal in the county of Shropshire, which only had the purpose of facilitating the transportation of iron ore and coal from Oakengates to the foundry at Ketley.*[28] (quoted from a travel diary[29] published at the beginning of the nineteenth century). The whole canal was only about $1^1/_2$ miles (2.5 km) long with a rise of 73 feet (22.25 metres) from Ketley. Its builder was William Reynolds, who owned the Ketley foundry.

The incline was built for twin operation, with one loaded boat descending hauling up an empty one. There was a lock at the summit of each incline, over which was a large wooden drum, around which ran a long rope. This was connected to the carriage on which the boats sat and the speed of operation was controlled by a brake on the drum.

The boats were box-shaped, 20 feet (6.10 metres) long, 6.33 feet (1.93 metres) wide and 3.8 feet (1.17 metres) draught, which allowed them to carry about eight tons. On the incline, they sat on a wagon, which had two pairs of wheels, one pair about $2^1/_2$ feet (0.76 metres)

diameter, and the other $1^1/_4$ feet (0.38 metres) diameter. The wheels had double flanges which fitted on to rails spiked to longitudinal sleepers, and cross-ties kept the rails parallel. The railway extended into the water so that a boat could be floated over the wagon. The slope of the incline was probably between 1:2 and 1:2.5, making it almost impossible to use larger boats, carried longitudinally, on the incline. A wagon for such boats would have had to be much longer and because of its height at its outer end would probably have been unstable. The tracks on the incline were about 7 feet (2.20 metres) apart and their gauge was about 6 feet (1.90 metres). On the canal, the boats were assembled into a train, and towed to their destination by horses.

On arrival at the top of the incline, the boat entered a lock chamber, which was only slightly larger than the boat. At the lower end of the lock was a counterweighted gate which could be lifted vertically. When it opened, the gate and counterweight were high enough for a boat to pass underneath. A conventional lock gate was fitted to the upper end of the lock, with a gate paddle to allow the chamber to be filled. Water

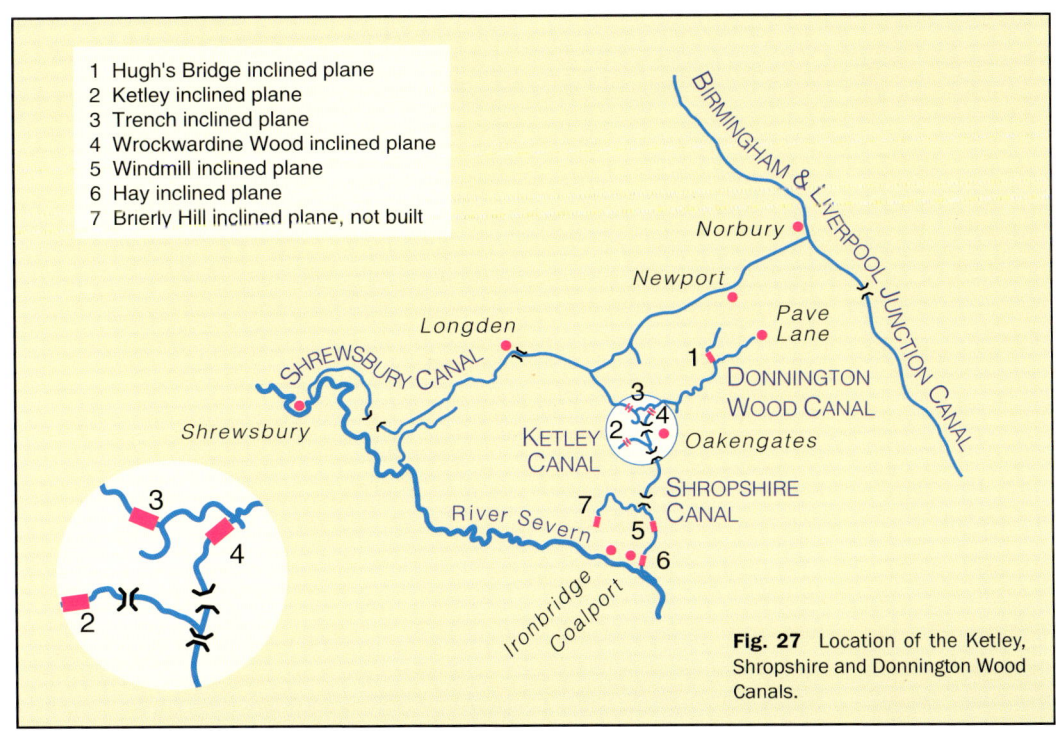

1 Hugh's Bridge inclined plane
2 Ketley inclined plane
3 Trench inclined plane
4 Wrockwardine Wood inclined plane
5 Windmill inclined plane
6 Hay inclined plane
7 Brierly Hill inclined plane, not built

Fig. 27 Location of the Ketley, Shropshire and Donnington Wood Canals.

drained through sluices in the lock wall into a cavity between the lock chambers, from where it ran along a channel in the lock foundations into a large basin about 15 feet (4.70 metres) below canal level. It was returned to the canal from here by a small steam pump.

The incline seems to have given complete satisfaction. On the 16th May, 1789, Reynolds wrote to James Watt: *Our inclined plane answers my most sanguine expectations. We have already let down more than forty boats a day each carrying eight tons — on an average thirty boats a day and have not yet had an accident.* In 1792 a copper token was minted which included a view of the brakeman at work.

After the Napoleonic war ended, the works and incline were closed. It was last used in 1816[30] and it is now difficult to find its location as the lower basin, which provided a reference point, was filled up in 1965.

Hagen[31] describes the incline, partly based on Dutens[32] description of the Shropshire Canal inclines, as follows: *Fig. 359a shows a cross-section through the upper part of the first inclined plane. The boats used here are very small and box-shaped without a chamfer on one side or the other. They are, as previously mentioned, directly fastened to each other, so that they form a long boat when using the canal. They are 19 feet long, near 6 feet broad and 3 feet high, and will carry 100 Centner [circa 5 tons], when they draw about 2 feet. The wagons, on which they are carried, are equipped with four wheels, of which one pair is 2 ¹/₂ feet diameter and the other only 1¹/₄ feet. Over the first section, the inclined plane is walled and at either end its slope is slightly flatter, so that the boat remains horizontal as it is lifted from the water or is put back again. On the long sides of the wagons there are light wooden bracings, strengthened by iron bars, which form part of the side walls and partly serve to protect the rope by which the wagons are pulled up and lowered. For this purpose, struts on either side are joined above by a bar, as shown in the cross*

section of the wagon, Fig. 359b. To this bar, two chains are fixed, with a hook at their ends, for strengthening the winding rope and each boat is provided at front and back with a hook onto which rings at the ends of the chains are fitted. If a boat entering the lock on the upper canal or over the wagon in the lower canal is so deep in the water that only the upper bar and the side ropes project from it, then the boat is pulled into the entrance and the chain is fastened first to the front hook of the boat, which prevents it from moving too far forward. Afterwards the ring of the second chain is hung on the boat's stern hook. This operation presents no problems with regard to the chains, in that they remain slack so long as the boat is floating and only tighten when the boat rests on the wagon. The gap between the guide walls and the boat's sides is so small, that when the water is let off the boat aligns itself automatically and precisely on the wagon. ...

Each lock chamber is just wide enough to take the wagon. Because the track continues over the lock floor, the wagon can enter and leave every time with safety and without touching the walls. The protection against the upper canal water is a simple lock gate, which blocks the small entrance. The figure shows it closed and the chamber empty. There is a sluice to fill the chamber.

The upper chamber of the incline is closed by a gate, which is shown opened. It hangs by two chains, which are wrapped twice around a wooden axle, and across their other ends is hung a prismatic piece of cast iron, C, which is a counterweight to the gate. When the gate is lifted and the counterweight lowers, they are both high enough for the wagon to pass under them. The operating rope fastened to the wagon does not prevent the gate from closing as it is only lowered when the wagon is in the chamber. The gate is adjusted such that it will lower automatically with a small additional weight. At the end of the lifting axle is a drum, H, around which a second rope is wrapped, and

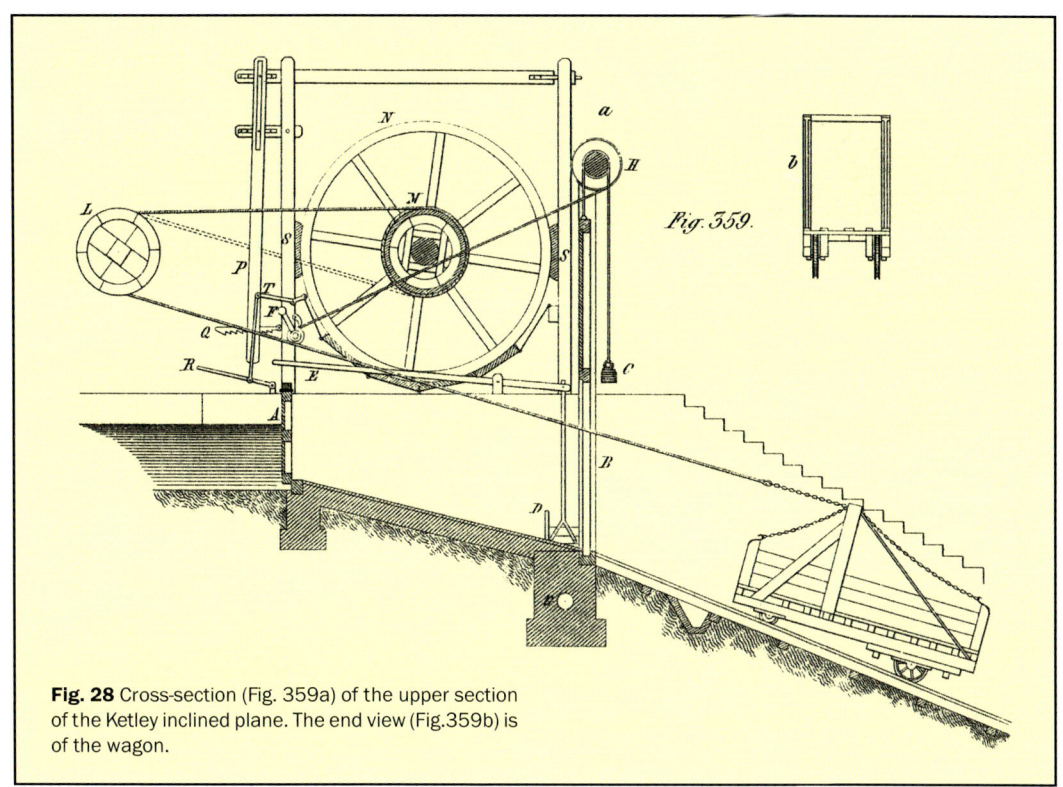

Fig. 28 Cross-section (Fig. 359a) of the upper section of the Ketley inclined plane. The end view (Fig.359b) is of the wagon.

this winds around a smaller drum on the same axis as crank F, which is seen alongside a ratchet. By means of this crank, the gate is lifted and lowered.

Each operating rope, by which a wagon is pulled up and lowered, hangs over the middle of the track and lock. When it is fully extended down the incline, there are cross-rollers along the track at about 15 foot intervals to prevent abrasion. This rope is fastened by means of a short chain to the bar of the frame on the wagon and is wound over a large wheel, L, to pull the wagon into the lock. This wheel is used as a simple solution to the problem of placing the main drum of the machine as far back as possible. The final drum in the figure, marked M, is located over the lock chamber and is constructed from wood planks. It extends over both chambers and the space in between. The hoisting ropes for both tracks are wound around it and their ends fastened to it. They are, however, wound in opposite directions, so that when the drum is turned, one rope is unwound

and the other is wound up. The solid line shows the front rope, the dotted line that for the other chamber. If the drum was left to itself, there would be a strong tendency for the speed of the loaded boat to increase as it descended and similarly the unloaded one ascending. To control such movement, there is in the middle of the drum a large brake wheel, N. It can be slowed partly by the frame which encloses it pushing on the lever P which holds the brake-blocks S against the outside of the wheel. There is also a brake chain on which wood is fixed under the wheel and which is operated by the lever T.

These different arrangements are so laid out, that the attendant can set each part in motion without having to move far. As soon as the wagon has been drawn into the lock chamber from the lower canal, the attendant lowers the gate by means of the crank F. Next, the worker standing on the boat opens the sluice in the upper gate, opens the gate as soon as the chamber is filled, and by means of a shaft releases the boat from the wagon and pushes it into the upper canal. A loaded boat is guided over the wagon and at the same time an empty boat is guided over the wagon in the lower canal. A bell, whose ring can be heard down at the lower canal, gives the signal that everything should be ready. Then the attendant steps onto the lever E and opens the sluice which empties the lock chamber. As this is happening, he presses on the lever P to prevent the wagon running away as soon as the boat rests on it. He puts the brake on firmly by engaging lever P with the notched iron bar Q. After this he hoists the large gate with crank F, opening the lock chamber to the inclined plane. Because the track, which lies within the lock, is only slightly inclined, the wagon will not move until the brake is released. As soon as the wagon has left the lock the brake must be applied again. Even then this does not prevent the speed from sometimes increasing too much. Then the machine

attendant steps on lever R which, through the angle lever T, applies the brakes. In this way the operation can be completely regulated. The time for raising or lowering a wagon is between two and three minutes.

The loaded wagon loses a part of its weight as it enters the water. At the same time the pull of the empty wagon, because of the lesser slope of the track on its entering the lock, also becomes somewhat less, so that the former can not run deep enough into the water for the boat to float free. As soon as possible a horse is attached and draws the wagon forward so that the boat floats freely and the next empty boat can be pulled in and fastened. An empty wagon is drawn in to the chamber in the same way should not have completely entered. However, for this purpose a special device is used. On the axle L is a gearwheel meshed with a pair of gears which can be turned by means of a crank. By this crank and by some workers grasping the spokes of the large

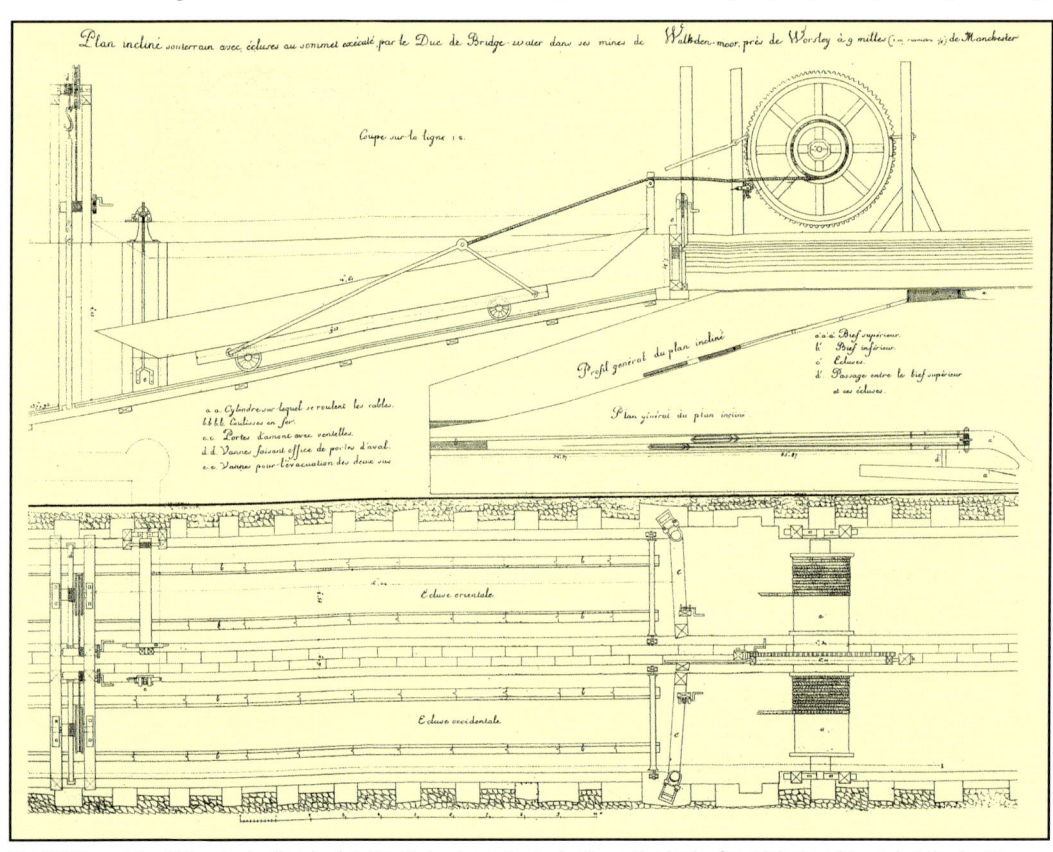

Fig.29 The original Worsley incline had a simple horizontal winch above the locks for raising and lowering the boats.

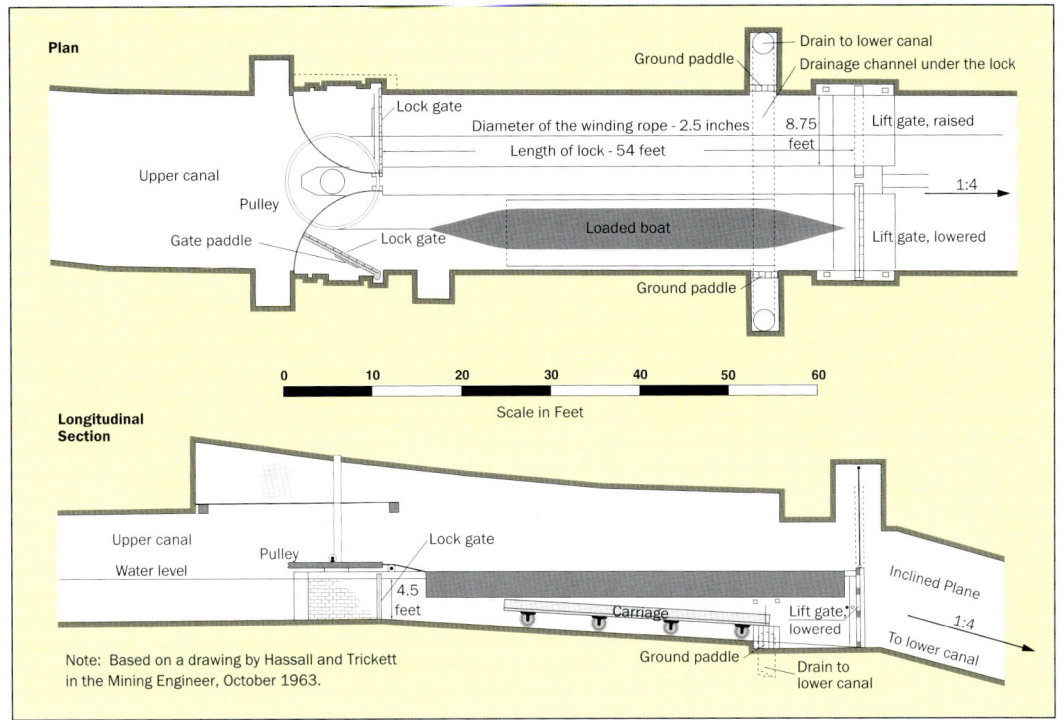

Plan

Ground paddle
Drain to lower canal
Drainage channel under the lock

Lock gate
Diameter of the winding rope - 2.5 inches
Length of lock - 54 feet
8.75 feet
Lift gate, raised

Upper canal

Pulley

1:4

Gate paddle
Lock gate
Loaded boat
Lift gate, lowered

Ground paddle

0   10   20   30   40   50   60

Scale in Feet

Longitudinal Section

Upper canal
Water level
Pulley
Lock gate
4.5 feet
Carriage
Lift gate, lowered
Inclined Plane
1:4
To lower canal

Ground paddle
Drain to lower canal

Note: Based on a drawing by Hassall and Trickett in the Mining Engineer, October 1963.

Fig. 30 Boats loaded with coal descended the incline, raising empty ones. The mechanism may have been rebuilt at some stage to include a horizontal pulley in place of the winch, as shown here. Only boats carrying maintenance materials needed to be hauled by a winch up the incline, which was usually self-acting.

brake wheel and turning it, the wagon and empty boat can be drawn up, even if no loaded one descends. This situation is unusual.

There may have been a second incline in the area as a plan from 1788 suggests that there was an underground inclined plane in a coal mine belonging to the Earl of Gower in the Donnington Wood area.[33] Unfortunately there are no further details, though John Gilbert was involved with mining here, and he was also involved with the underground incline in the mines associated with the Bridgewater Canal at Worsley.

Worsley, the underground incline

The Bridgewater Canal, the first section of which opened in 1765 from the coal mines at Worsley to Manchester, was built by Francis Egerton, the 3rd Duke of Bridgewater. What is today the main line of the canal opened in 1776, and runs from Manchester to Runcorn. It is 28 miles (45 km) long and had a flight of locks to the Mersey at Runcorn now filled in. There are two branches; to the Trent & Mersey Canal at Dutton and the 10 mile long Leigh branch. The

first section of the latter, connecting with the underground mine system at Worsley, was part of the original canal. The mines were the Duke's property and were first drained around 1730 by an adit which led from within the mine to Worsley Brook. John Gilbert, the Duke's agent, had the idea to build a new, improved, drainage system such that boats could use it, and in 1765 this idea was realised. By 1840, over 46 miles (74 km) of underground canals had been built. The main canal level was four miles (6.4 km) long and the upper level $1^3/_4$ miles (2.8 km). There were two deeper levels, the 46 miles also including numerous branch canals. The main underground canal was around 10 to 12 feet broad (3.05-3.66 m), with 8 feet (2.44 m) clearance above water level and a water depth of up to 4 feet (1.22 m). Part of the main system was lined with bricks, particularly where the branches joined the main canal. Shortly before his death, Gilbert sketched a design for an inclined plane for boats which would join the main canal with the upper level.

Fig. 31 The top of the incline around 1960 after one incline had been filled with silt removed to improve drainage.

Robert Fulton was involved with the trial of a steam boat in Manchester at this time. His ideas for boat lifts were published in 1796 under the title *A Treatise On the Improvement of Canal Navigation, exhibiting the numerous advantages to be derived from small canals and boats of 2 to 5 feet wide, containing from 2 to 5 tons burthen,* so he may have had some influence on the design. The original coal handling system, baskets raised and lowered in vertical shafts, was replaced by the new incline.

Construction of the incline, which was 453 feet (138 metres) long and raised boats $106^1/_2$ feet (32.46 metres), began in September 1795, and was completed by October 1797 at a cost of £1,781-5s-1d (£1,781.25). At the upper end two locks were constructed from which two railway tracks ran to the lower level. The locks had brick walls, with separate lock gates at each end and the two tracks were brought together 57 yards (52.1 metres) from the lower end of the inclined plane. The boats, maximum size 55 feet (16.76 m) by $4^1/_2$ feet (1.37 m) and 12 tons load, were carried on wagons 30 feet (9.14 metres) long and 7 feet (2.13 metres) broad. The incline was operated as a twin lift, the weight of the descending loaded boat being sufficient to raise an empty one. To allow loaded boats to ascend and to control their ascent and descent a horizontal winding drum was fitted. From an inspection of the remains carried out around 1960, this may have been replaced by a vertical drum with a diameter of $15^1/_2$ feet (4.72 metres) in the 21 feet (6.40 metre) high 'cave' over the locks. The speed of the winding drum was controlled by a brake and in eight hours about 30 boats could be handled.

Operating the lock at the upper end of the inclined plane proved time-consuming and caused a large loss of water. Also, the unpowered operation by counter-balance was uncertain and ways were sought to improve these details as much as possible on later inclines.[34]

The incline worked until 1822 and the main underground canal until 1887. After closure, the canal continued in use for drainage and was regularly inspected by boat until the 1960s. It was finally closed in 1968. Today you can still see the entrances to this vast tunnel system from the road in Worsley. Boats cannot reach them from the waters of the Bridgewater Canal because of silt, though the first section of the tunnel was cleared and inspected recently to ascertain its condition and it may be reopened in the future as a tourist attraction.

Fig.32 The original entrance to the Worsley mines.

Fig.33 The Hay inclined plane at the end of the nineteenth century.

The Shropshire Canal inclines

In 1788 the construction of the Shropshire Canal (see map on page 23) began under the direction of William Reynolds. It ran for about 12.5 km from the Donnington Wood Canal to the River Severn. As well as two approximately 825 feet (250 metres) long tunnels there were three inclined planes, at Wrockwardine Wood (or Donnington Wood), Windmill Farm and The Hay, Coalport. The rises were respectively 120 feet (36.57 metres), 126 feet (38.4 metres) and 207 feet (63.09 metres). The last was only exceeded in 1819 with the construction of the 225 foot high inclined plane at Hobbacott Down on the Bude Canal. The Hay incline lies within the Ironbridge Gorge Museum, north-west of Birmingham, one of England's largest and most interesting museum complexes. The museum is named after the world's first cast iron bridge which was erected by Abraham Darby III in 1777-1779 and which spans the River Severn at Coalbrookdale in a bold arch. The incline is a couple of hundred metres downstream on the northern bank of the Severn. The remains comprise railway track installed circa 1975, the dry summit about one foot (30 cm) above the water level of the upper canal (the canal level has changed due to subsidence), the upper canal works and masonry foundations from the former machine house in which the steam-engine worked.

The Hay, as well as the other inclines on the Shropshire Canal, worked as a twin lift. Usually descending loaded boats hauled up empty ascending ones. On all the inclines the summit was above the water level of the upper canal so that locks were not needed. The tracks were laid over the summit, continuing downwards on each side. Originally boats were hauled over the summit by man-power or horse-power, but this was changed in the early 1790s to a steam-engine.

The Prussian engineers von Oeynhausen and von Deschen give an excellent description of The Hay incline which was published after a study trip in 1826/27 to look at industry in England.[35] Their description of the *'Bremsberg from the Shropshire Canal to the Severn, not far from Coalbrookdale'* is as follows:

About two miles below Coalbrookdale, the Shropshire Canal is united with the Severn by a self-acting inclined plane 793 ft long and with a vertical drop of 207 ft, or an angle of fifteen degrees. The railway thereon is of cast-iron, built in the manner of a tramroad. The rails are the strongest and thickest of the kind that we have seen; also, they lie on longitudinal timbers 14 inches wide and thick, and these sit upon

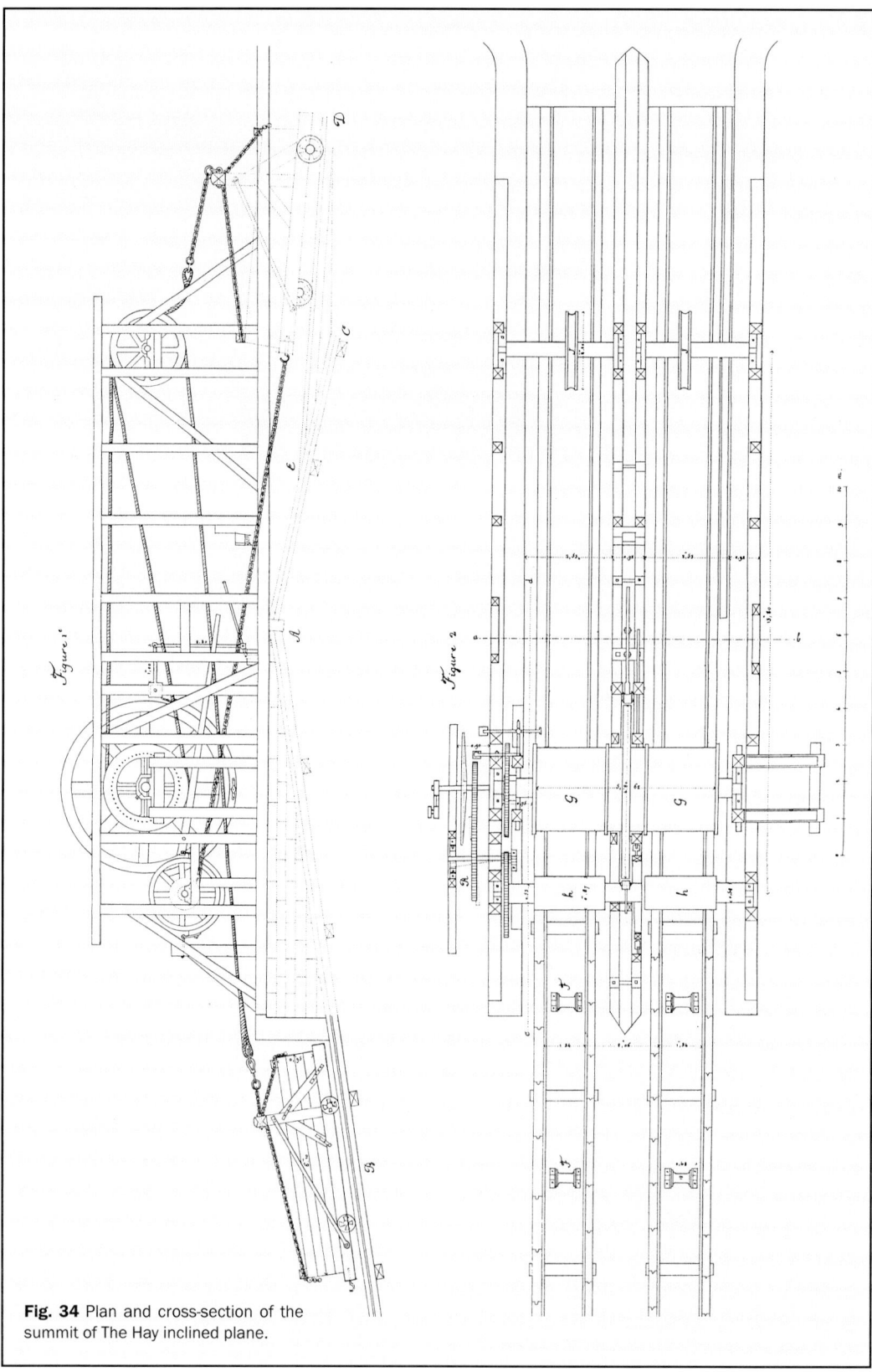

Fig. 34 Plan and cross-section of the summit of The Hay inclined plane.

Fig. 35 A view towards the summit of The Hay inclined plane, with the remains of the steam engine house on the right.

wooden cross-sleepers. The whole incline is well paved. The tramroad has three lines of rails, so that it forms two tracks with one common rail; in the middle of the plane, where the full and empty wagons pass one another, there are four lines of rails forming two separate tracks. On one track the rail flanges are inside, and on the other track outside. The rails are 70 inches long; 7 inches wide, exclusive of the $2\frac{1}{2}$ inches high and 1 inch thick flange; and 2 inch thick. On one end there is a projecting point 5 inches broad and $1\frac{1}{2}$ inches long and on the other end a suitable notch into which the point fits. In addition, there is a tab on the flange through which a wooden plug is inserted to connect the two rails; four 1-inch square holes enable the rails to be spiked to the longitudinal timbers.

The connection of the rails by wooden plugs is seldom used and only a few rails are provided with the small recess in the middle of the flange. In the track are iron rollers over which the rope runs.

The canal boats arriving by the Shropshire Canal are let down on this self-acting plane. A wooden wagon frame is used for lowering them. Most of the canal boats are of wood, 18 ft long, 5 ft 2 inch wide, and $2\frac{1}{2}$ ft deep, in the clear. One such boat weighs about $1\frac{1}{2}$ tons empty, and the greatest loads are about 5 tons, whereby it still has about 3 in. freeboard. The cargo consists of coal and iron.

The wooden frame on which the boats are placed is of very simple construction. On the two outer frame beams stand posts 5 ft high, fixed by struts, and connected at the top by an iron cramp provided with three eyes for the attachment of the chain. The frame has two large front wheels 27 in. diameter, and two small hind wheels 16 in. diameter, which have an inside gauge of 43 in. They are 6 in. wide at the rim, and 10 in. long in the nave; these lie wholly under the frame. Outside of the wagon there are also two outer flanged wheels of 24 in. diameter on the same axle as the small hind wheels, and with a gauge of 78 in. When the wagon is drawn out of the canal upon the bank of the self-acting plane, these wheels run upon a special railway, and thereby enable it to pass over the uppermost point of the bank without rubbing. The weight of the wagon frame is 2 tons; and the boat with its load $6\frac{1}{2}$ tons; thus the total weight is $8\frac{1}{2}$ tons or 170 cwt. About 100 boats can be let down in 12 hours.

A 16 inch steam-engine stands on the highest point of the self-acting plane, on the summit formed by it and by the short inclined plane leading to the canal. This serves to draw the wagon with the canal boat out of the upper part of the canal. It also completes the drawing up of the wagon coming up the self-acting plane, when the full wagon with the boat has plunged

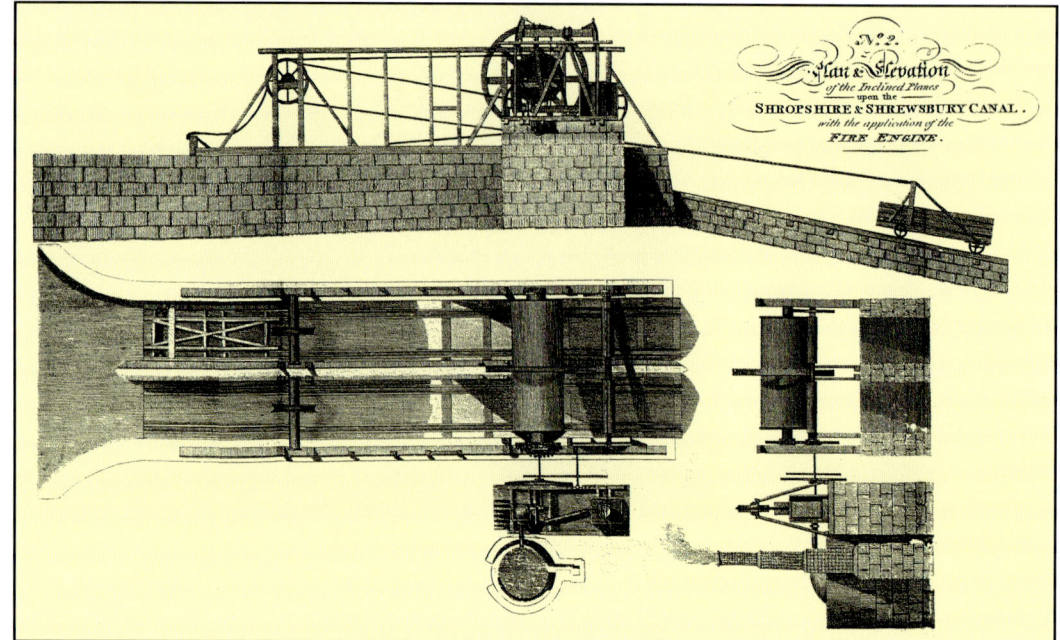

Fig. 36 A contemporary drawing of a Shropshire Canal inclined plane, showing the steam engine.

into the water and thus so reduced its effective weight that it is no longer able to draw the empty wagon right to the top of the plane.

On the axle of the steam-engine there is a small gear-wheel with 24 teeth, which engages with a larger wheel with 96 teeth, on the axle of

Fig. 37 A recent view of The Hay inclined plane.

which is fitted the rope drum of 7 feet diameter, with the brake wheels for the self-acting plane. The small gear-wheel can be put in or out of gear by a contrivance operated in the engine house, so that the engine and the rope drum can move independently of one another. The rope goes from this drum over iron rope-sheaves 6 feet diameter and 5 inches wide and then under the drum away to the self-acting plane; the rope-sheaves hang perpendicularly over the upper part of the canal, so that the wagon can be drawn forward over the ridge.

A pinion with 40 teeth on the axle of the steam engine engages with a wheel with 96 teeth mounted on an intermediate axle which also carries a pinion with 28 teeth. The intermediate axle has a bearing-seat inside the engine house and can be put in or out of gear by means of a lever, so that this axle can be put in motion or at rest according to choice. The pinion of 28 teeth engages with a wheel of 80 teeth which is mounted on the axle of a small chain-drum and brake wheel; through this drum, and a chain, the canal boat is drawn out of the upper part of the canal on to the top of the self-acting plane. The operation of the brake is as follows — As

Fig. 38 The author at the summit of The Hay incline.

soon as the full wagon arrives at the bottom, the laden boat is removed by a workman and pushed into the canal and, in exchange, an empty boat is brought on and affixed. The full down-going boat does not draw up the empty one the whole distance because the tramroad goes entirely under the surface in the lower canal and the loaded boat as soon as it plunges into the water loses much of its weight. As soon as the empty boat stops, a workman releases the brake on the large rope drum, while another puts the engine in connection with this drum and sets it in motion so that the boat shall continue its journey upwards and across the ridge of the self-acting plane. When the boat has reached the ridge it goes slowly on the tramroad into the upper canal. The large rope-drum is then disconnected from the engine, the engine stopped, and two workmen push the empty boat off the wagon, put on a full one instead, and attach the wagon to the chain of the small drum. This chain was unwound during the time that the empty boat was passing over the ridge, at which time also the chain-drum was connected with the engine. As soon as the full boat is pushed on the engine is started and, by means of the chain, draws the full boat upwards out of the canal on to the ridge of the self-acting plane. One workman stands at the engine, the other at the brake. As soon as the boat is across the ridge, it is attached to the rope of the large drum and let down the plane by means of the brake. To permit this, the engine is stopped and the

chain-drum disconnected from it; this done, the workman removes the chain from the wagon and puts on the brake, keeping it on until the full boat plunges into the water of the lower canal. In between, a workman has time to look after the firing of the engine, and to bring the boats into position and attach them, so that little time is lost. The workman at the lower canal is also employed in the unloading, etc., as affixing and detaching the boats requires little time. At Coalport on the lower canal a ton of coal costs 8s 0d, a very low price when the fact that already the coal has come from the pit many miles distant is taken into consideration.

All of the connecting wheels of the brake-mechanism are cast in two parts, which reduces the danger of breakage when they are being staked on their axles. The axles are of cast-iron 4 in. square, the journals $3^1/_2$ in. diameter, the wheels 3 in. in the split part. The engine crank is 18 in. long. The rope which lets down the wagons is $3^1/_2$ in. diameter.

The large rope-drum makes a quarter of a revolution for one revolution of the engine shaft and the small chain-drum makes $^7/_{48}$ or about $^1/_7$ of a revolution. Tredgold, in his book: A Practical Treatise on Rail Roads and Carriages, gives the length of this incline as 1,050 ft. In addition, he states that the Shropshire Canal has two other self-acting planes; the second 1,800 ft long and with 126 ft fall, or an angle of just over 4 degrees; the third 1,050 ft long and with 120 ft fall, or an angle of $7^1/_6$ degrees.

Fig. 39 The summit of the Trench inclined plane showing the engine house and a tub-boat about to descend.

The inclines on the Shropshire Canal came into service in 1791. Around 1858 part of the canal was closed together with the planes at Wrockwardine Wood and Windmill Farm. The canal section on which The Hay was situated finally closed in 1894, so the incline remained in service for over 100 years.

A good section of the canal survives in the open-air museum, as well as two of the typical tub-boats, 18 feet (5.49 metres) long, 5 feet (1.52 metres) broad and loading 5 tons.

Hugh's Bridge and Trench inclines

In connection with the operation of the Shropshire Canal, two more inclined planes were constructed. One was on the Donnington Wood Canal at Hugh's Bridge and was in service from around 1790. It had a rise of some 36 feet (11 metres) and had probably closed by 1879. Better-known is Trench incline on the Shrewsbury Canal. Both these inclines (see map on page 22) were constructed in the same way as those on the Shropshire Canal, with a dry summit. They were twin inclines whose operation was aided by steam-engines and were used by typical Shropshire Canal tub-boats capable of carrying about 5 tons. The Trench inclined plane was particularly noteworthy. It entered service in 1793 and was in use for well over 100 years. It was the last incline in use in Britain, only ceasing operation on the 31st August 1921.

Fig. 40 The Trench inclined plane early in the twentieth century.

The incline was 670 feet (204 metres) long with a rise of almost 75 feet (23 metres). It could transport ten tub-boats an hour but the average in good years was around 50 to 60 boats per day. Up to the end of the First World War the main traffic was coal descending and wheat ascending for a mill at Donnington. Until recently you could still find remains of the incline, though now much of the site has been destroyed by road building. The Shrewsbury Canal formed the link between the local tub-boat canals and the main English canal system via the Newport branch of the Shropshire Union Canal. Immediately below the Trench incline the locks were 81.5 feet (24.8 metres) long and 6.5 feet (2 metres) wide, and so were too narrow for typical narrow-boats.

The South Hadley Canal

In the eighteenth century there were efforts outside Great Britain to introduce inclined planes. These included one built in the USA in the early 1790s to link the South Hadley Canal with the Connecticut River on which boats up to 65 feet (20 metres) in length carrying up to 25 tons could be moved. It overcame a rise of about 53 feet (16 metres) with a lock at its upper end. There was a single carriage for carrying boats, powered by a water wheel. The incline was used mainly for the transport of wood and was abandoned by 1805 in favour of locks.

Fig. 41 Model of an inclined plane designed for use on the Unstrut Navigation.

The Klodnice Canal inclines

Prussia also saw the introduction of the inclined plane. There is a model of one in the Berlin Verkehr und Baumuseum whose design was proposed for weirs on the River Unstrut between 1791 and 1795, possibly at Ritteburg, Schönewerda and Nebra. A descending carriage with a loaded boat hauled up an empty boat through a return pulley so it could only be used

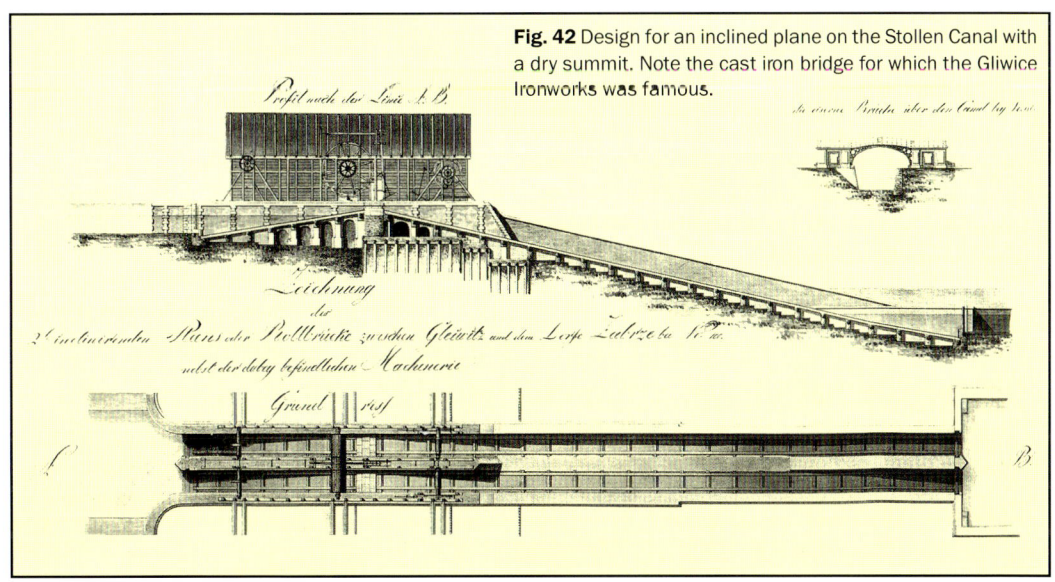

Fig. 42 Design for an inclined plane on the Stollen Canal with a dry summit. Note the cast iron bridge for which the Gliwice Ironworks was famous.

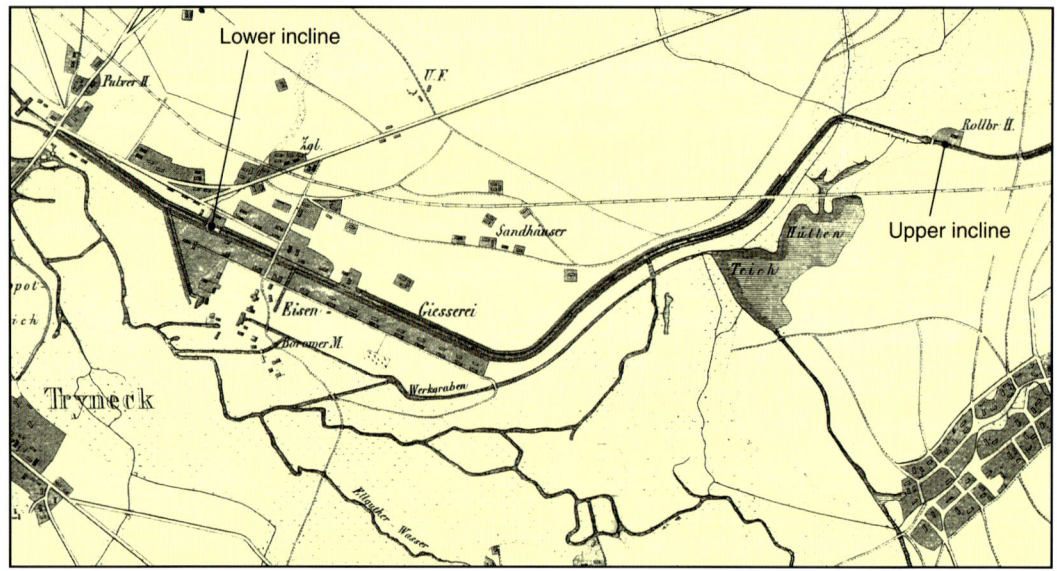

Fig. 43 Map showing the location of the inclined planes on the Stollen Canal; the upper incline is marked as 'Rollbrücke'. The Royal Ironworks is indicated by 'Giesserei', and the coal mine was off the map to the right.

for the downstream movement of goods. It is unlikely that these inclined planes were built and probably 'Kunstmechanikus' Mende, who built Germany's first vertical lift in Halsbrücke (see page 83) and who was responsible at the time for making the Unstrut navigable, had suggested the use of inclines but the idea was abandoned. Chamber locks were probably erected at the weirs from the beginning.

The first inclined planes in Prussia were built between 1801 and 1806 on the 8 km long Stollen (Mine) Canal at Gliwice (Gleiwitz) in Upper Silesia. It formed an extension to the Klodnice (Klodnitz) Canal (now replaced by the Gliwice Canal) which ran from the Oder to the Royal Ironworks in Gliwice. The extension linked coal

mines to the ironworks and could be used by boats 6.26-8.60 metres long, 2.04 metres broad and of 4 tons capacity. To overcome the change in levels, two inclined planes were built with rises of 11.60 metres and 5.02 metres. The original design was based on the inclines in Shropshire and boats were to be pulled out of the water and over a summit before descending. It was changed after news of the underground incline at Worsley reached Prussia and in the Verkehr und Baumuseum, Berlin, there is a wooden model of *'the inclined plane for the tunnel canal'*. It was presented by the regional government office in Opole (Oppeln) in 1908 and shows that a coal boat resting on a carriage was not pulled over a summit, but was placed

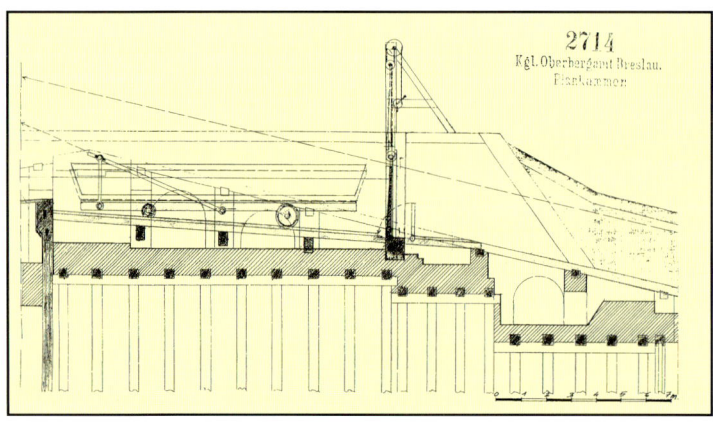

Fig. 44 A cross-section of the lock at the top of the Klodnice Canal inclines. It was an improvement on that at Worsley on which they were based. Note the hinged section of track which ensures that the gate is water-tight.

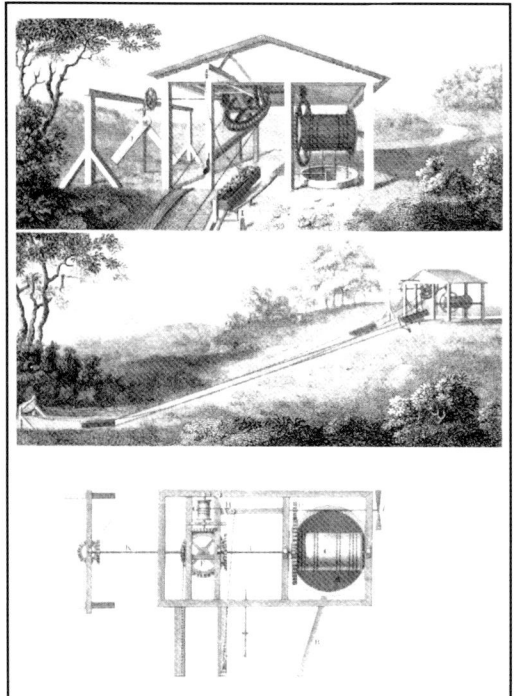

Fig. 45 Fulton's design for a bucket-operated inclined plane.

on the carriage in a lock chamber at the upper canal level — just like at Worsley. This model is similar to plans from the drawing office of the Royal Mining Office at Breslau.[36] *The lock was 13³/₄ metres long and 4¹/₂ metres wide. As loaded boats descended and empty ones ascended, there was no powered drive… In 1824 the lower incline*

was taken out of service, followed in 1839 by the upper one. The remains gave no clue to its former condition.[37]

Robert Fulton and the Canal du Creusot

Elsewhere, between 1790 and 1800, there were numerous new ideas for improvements not just to inclined planes but also for vertical lifts. Many did not survive the experimental stage but some were introduced successfully later.

The great American engineer, Robert Fulton (1765-1815), who built the first practical steam ship, the ***Clermont***, in 1807, obtained a patent on 8th May 1794 for '*a mechanism for moving canal boats from one level to another*'. Two years later he added to the ideas proposed in this patent for inclined planes in an essay about improvements to inland shipping.

He proposed fitting tub-boats with wheels such that they could be run up or down an incline. This idea was only suitable for small boats and was used 25 years later when the Bude Canal inclines were built. Earlier, the system had been tried in 1806 on the French Canal du Creusot, near Torcy (to the east of Paris), where three vertical lifts and three inclined planes, with rises of between 5.6 and 8.6 metres, were proposed for 8 ton boats. Only

Fig. 46 Fulton's design for an inclined plane powered by a waterwheel.

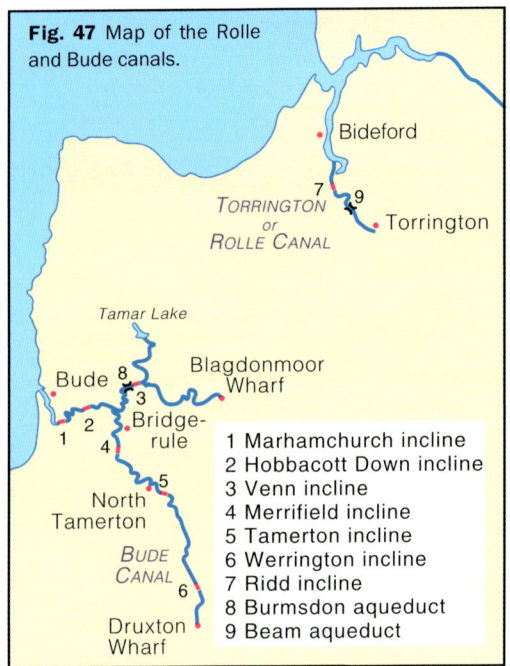

Fig. 47 Map of the Rolle and Bude canals.

Bideford

TORRINGTON
or
ROLLE CANAL

Torrington

Tamar Lake

Bude

Blagdonmoor
Wharf

Bridge-
rule

North
Tamerton

BUDE
CANAL

Druxton
Wharf

1 Marhamchurch incline
2 Hobbacott Down incline
3 Venn incline
4 Merrifield incline
5 Tamerton incline
6 Werrington incline
7 Ridd incline
8 Burmsdon aqueduct
9 Beam aqueduct

of the canal had been completed by 1823 and a branch to Druxton was opened in 1825. There were six inclines, three on the Holsworthy Branch at Marhamchurch, Hobbacott Down and Venn or Vealand and three on the Druxton Branch at Merrifield, Tamerton and Werrington or Bridgetown. Hobbacott Down incline had a rise of 225 feet (68.58 metres) which was only exceeded in 1968 with the construction of the Krasnoyarsk incline in Russia. Marhamchurch incline had a rise of 120 feet (36.58 metres), though the other four were smaller with rises of 51-60 feet (15.54-18.29 metres).

In constructing these inclines, Green had put into action two of Fulton's ideas. One was to provide the tub-boats with wheels, diameter 14 inches (35.6 cm), so that they could be drawn along the inclines. The tub-boats were 20 feet (6.1 metres) long, 5.5 feet (1.68 metres) broad and carried about 5 tons. They were towed in trains of four or five boats and the foremost boat was provided with a pointed bow.

While five of the inclines were powered by water wheels, Green introduced a second of Fulton's ideas at Hobbacott Down, where the incline was worked by means of the bucket and well system. There were two wells at the upper end of the incline as deep as the level of the lower canal, one for each track as the incline was designed for twin operation. There was a huge bucket of 10 feet (3 metres) diameter and $5^{1}/_{2}$ feet (1.68 metres) height in each shaft which

one vertical lift and an inclined plane were built, thus realising Fulton's ideas, though neither got beyond trial operation. The incline appears to have operated satisfactorily, but with large water losses. Among suggestions for their improvement was the first proposal for carrying boats in a water-filled tank or caisson. However, such a scheme was not yet practical.

The Bude Canal

The Bude Canal started at the port of Bude and had a total length of 57 km. It was begun in 1819 under the direction of James Green, a name that will appear again later. The main part

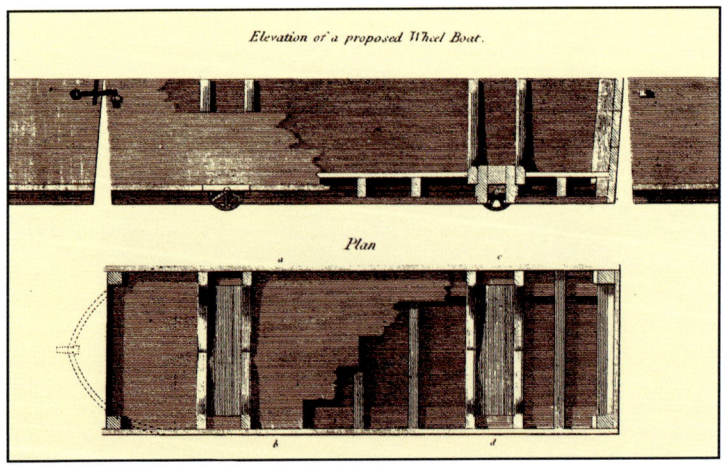

Elevation of a proposed Wheel Boat.

Plan

Fig. 48 A wheeled tub-boat as used on the Bude and Rolle canals.

Fig. 49 An engraving of the Rolle Canal showing tub-boats being towed by horse near Beam Aqueduct.

could hold about 13.5 tons, of water. They hung from a chain which ran over a drum and over which an endless chain to which the boats were attached also ran. The descent of the water-filled bucket in the shaft powered the incline. On reaching the bottom, the water in the bucket was automatically discharged into the lower canal. One operation took just five minutes.

All six inclines suffered from a variety of problems. At Hobbacott Down, the buckets and drive chains were said to have broken regularly so that a steam-engine was kept in reserve. Despite this, the bucket system was in service until 1891 when all the inclines and the tub-boat canal were closed. The wide canal between Marhamchurch and Bude continued to be used by larger canal boats. About two kilometres of this section can still be used today and the slope of some of the inclines is visible.

The Rolle Canal

A further inclined plane was constructed by Green in 1827 on the Torrington, or Lord Rolle, Canal at Weare Giffard. Six miles long and lying north-east of the Bude Canal in Devon, it was the private canal of Lord Rolle. The incline was planned as a twin lift which was to be powered by a water wheel, but little else is known. The boats mainly carried limestone and coal downwards and agricultural products upwards and were also to have wheels. The canal closed in 1871.

The Tavistock Canal

In 1819 at the same time as the Bude Canal was being built, an incline with a rise of just 19.5 feet (5.95 metres) was being constructed on the Millhill branch of the Tavistock Canal, deemed necessary because of the scarcity of water in the area. The boats were 30 feet (9.14 metres) long, 4.5 feet (1.37 metres) broad and 2.5 feet (0.76 metres) draught and carried $4^1/_2$ tons. They were hauled over the single-track incline by horses. The incline did not last very long and it was probably closed between 1831 and 1844. Another incline was situated at the

Fig. 50 Location of the Tavistock Canal.

Tavistock
Gunnislake
TAMAR MANURE NAVIGATION
TAVISTOCK CANAL
Calstock
N
Plymouth
1 Mill Hill Branch Inclined Plane
2 Morwelldown Tunnel
3 Morwellham Inclined Plane
4 Cann Quarry Canal

39

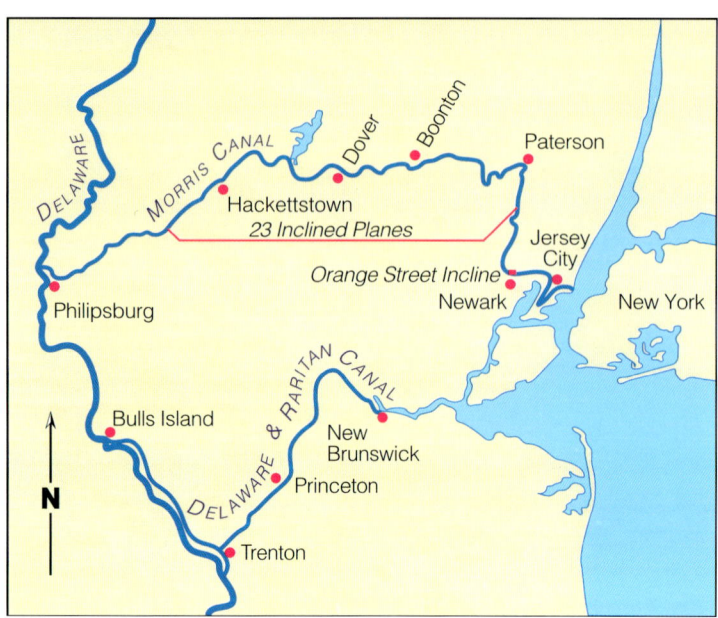

Fig. 51 Location of the Morris Canal.

lower end of the Tavistock Canal. This, however, was used by wagons which carried goods, mainly limestone, copper, slate and a little granite, transhipped from the boats. It lowered these materials some 240 feet (73 metres) to Morwellham Quay where cargoes could be transhipped to and from sea-going vessels.

The Morris Canal inclines

The 102 mile (163 km) long Morris Canal was built in the USA between 1825 and 1831, joining the Delaware River at Philipsburg with the Passaic River and thus to the Hudson River at Newark, opposite New York. It had 23 inclines (later 24), the greatest number on a single canal.

From the Hudson at Newark to its summit level, the canal rose about 914 feet (279 metres), then fell 760 feet (232 metres) to the Delaware River, at Easton. There were 16 locks and 12 inclines on the eastern end of the canal, with a further 7 locks and 11 inclines on the western end. The rises of the inclines, some single and some double track, varied between 35 feet (10.67 metres) at Port Delaware and 100 feet (30.48 metres) at Port Warren. Initially the boats using the canal could carry 25 tons, but this was raised to around 70 tons after the inclines were altered and the size of boats increased.

Since the inclines on the Morris Canal were later used as the model for the Elblanski (Elbing Oberland) Canal in Poland, this description by Hagen[38] is of interest and was based on an earlier account by Chevalier.[39]

Chevalier has described the double-tracked inclined plane which he saw at Philipsburg in 1835. It is the largest of all, with a rise of 97 feet. Its length is 89 Ruthen [about 1,100 feet], so it is inclined at about 1:11. Fig. 360a shows the upper part and the lock chambers which connect it to the canal. One chamber is empty and the other full. Fig. 360b is a cross-section of one lock and 360c of the other. Finally Fig. 361a and 361b show a carriage at the lower end of the incline and in cross-section in Fig. 361c.

The boats which travel on the canal are only modest in size and originally they carried not more than 500 Centner (25 tons). Later, by increasing the lock size, this was increased to 700 Centner (35 tons) and this has been further increased recently. Chevalier gives the lock size as $10^{1}/_{2}$ feet wide and 76 feet long, with the boats 10 feet broad and 60 feet long. Because the boats which use the locks have to rest on carriages when using the incline, so the lower end of the incline and its upper chambers must be of equal size. The carriages have eight wheels of equal size, like eight-wheeled railway wagons. Their axles are fitted rather close together in two pairs and are connected by a special frame or bolster which has, in its centre, the main bearing upon which the frame carrying the boat sits. The advantage of this is that the load is divided evenly between both axles and on each of the four wheels. The frame, Fig. 361, which sits on several strong transverse beams, is fitted on each side with trusses. Four columns are fitted over the bearing on the bolsters and these carry the load taken by the trusses. The columns support the middle of the frame and iron bars extend down from the outer end of each side

wall. Finally, three transverse beams connect the trusses and these are high enough so that boats can be floated below when they are loaded or unloaded.

In the middle of the floor of each lock there is a plank 'A' to which are fastened both ends of a chain which runs over three iron wheels. One, 'G', lies between the chambers and the others are on the lower ends of the carriages. When the carriage begins to descend, the chain is supported by light rollers as shown in Figs. 360 and 361. Motion is maintained by the drive wheel 'G'. This is 8 feet in diameter, is set at an angle, and is located centrally under both lock floors. A bevel gear is cast integrally with it and

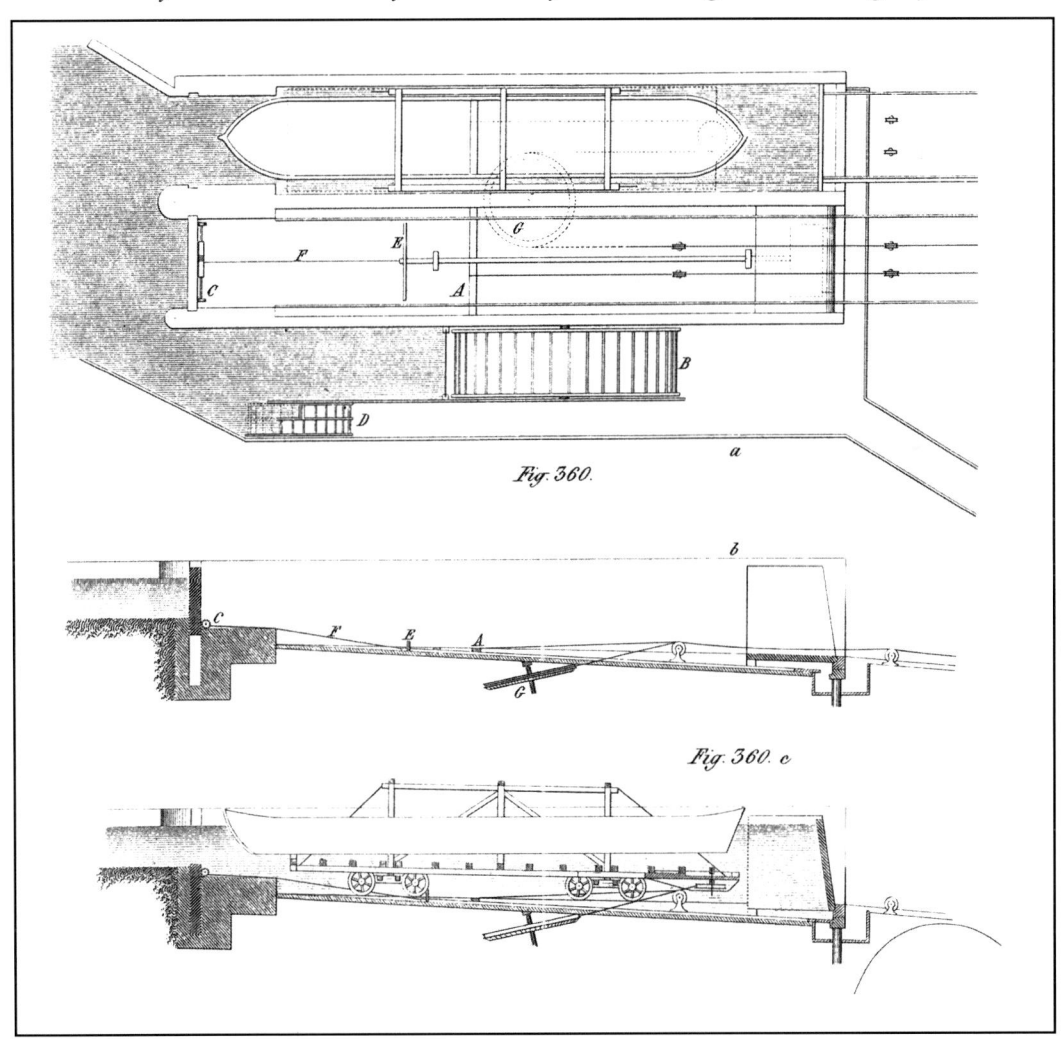

Fig. 360.

Fig. 360. c

Fig. 52 Ground plan (Fig. 360a) of the upper end of the Philipsburg incline. The cross-sections below show an empty lock (Fig. 360b) and a full lock (Fig. 360b).

41

this is driven by the large water wheel 'B' which is connected to it by gears. It is a high-breast water wheel and is fed from the upper canal. It is connected to the drive wheel by two bevel gears on a common shaft such that only one gear can be engaged at a time. In this way, the drive wheel can be turned in one direction or the other even though the water wheel always revolves in the same direction.

The gearing cannot be fitted closer to the lock chambers because the water loss at every filling would be too great. Chevalier does not mention a seal and this was probably provided by a gland on the drive shaft. An $8^1/_2$ Linien ($1^1/_4$ inches) chain was used and one must assume that it was made with specially shaped links which fitted firmly against the drive wheel 'G'.

The locks, which are built of wood, are side by side. At their lower end are single gates, which turned on a horizontal axis. When they are closed, as in Fig. 360 c, they are not vertical but are slightly inclined towards the chamber. The side wall rebate against which they seal also has to have this inclination. Nothing is required

to open them and they drop automatically as soon as the water level is low enough. However, they do not rest on the floor as they are fitted with rails on their reverse side and these must line up with the track. The gate at the upper end of the lock is opened not by lifting or turning, but sinks vertically so that boats can pass over it. Each gate is fitted with cast iron racks on its chamber side and two gears on a common shaft engage with them. These gears are driven by a second, smaller, water wheel 'D'. Because each gate must move independently and, of course, be either raised or lowered, the system used on the main water wheel was not suitable. Here the water wheel can be turned in opposite directions and is formed by two overshot water wheels side by side.

At the lower end of the chamber floors there is a large opening which leads into a channel below. This joins the tailrace of both water wheels from where the water falls into the lower canal or, if the water is not needed, is led away. The opening is covered by a horizontal sluice and is coupled by bars, which lie flat on the lock

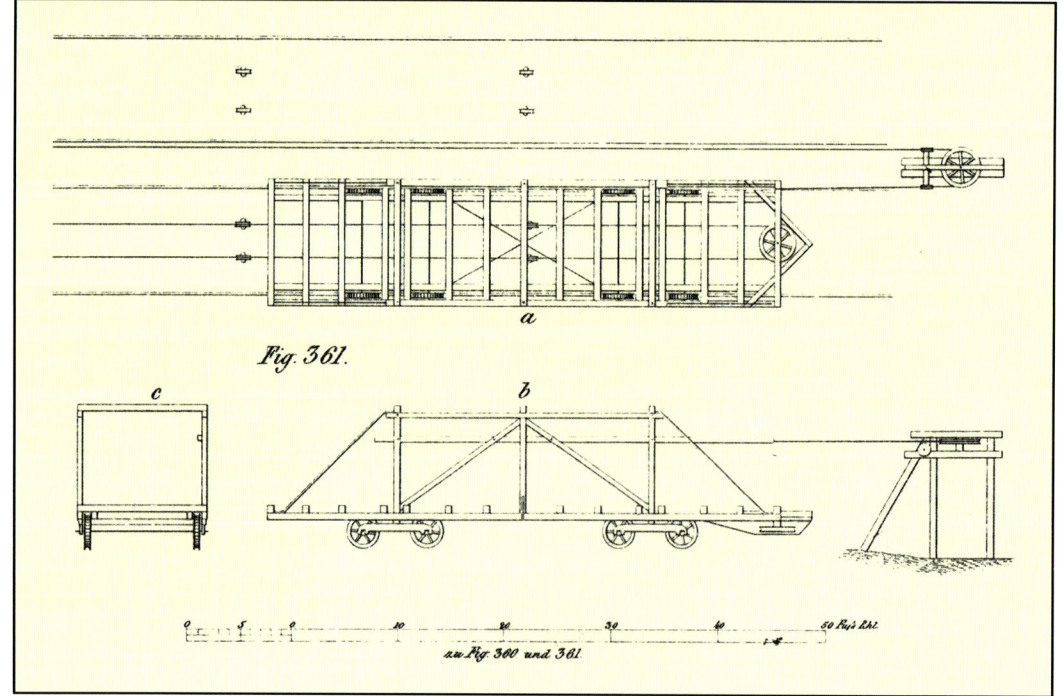

Fig. 53 Plan (Fig. 361a) and section (Fig. 361b) of the lower end of the Philipsburg incline. The carriage is shown in end-section (Fig. 361c).

Fig. 54 Inclined plane No.6, near Rockway, around 1900, with a sectional boat just leaving the carriage. In the background, to the left of the incline, is the turbine and machine house.

floors, to a plank 'E' with its edge uppermost. Chains 'F' are connected to the plank and these are wound around the previously mentioned shaft which closes the upper gate. The chains are adjusted so that as soon as the gate has almost completely risen and thus closes off the chamber from the upper canal, it then pulls open the horizontal sluice in the floor of the lock.

The chamber has now been emptied, the upper canal gate has been lifted and the lower gate has been lowered and the sluice opened to empty the chamber. All the parts are now as illustrated in Fig. 360 b and the chamber is ready to receive the carriage. This enters from the incline over the folded down lower gate and, after this, the attendant starts the smaller water wheel, turning the gears 'C' engaged with the toothed bar on the gate and lowering it. The

chain twisted around the gear shaft is unwound, freeing the slide bars fitted to the floor of the lock chamber which operate the horizontal sluice. Water from the upper canal begins to flow over the upper gate into the lock chamber and runs down the inclined floor of the chamber (its slope is considerably steeper than the track, as shown in Fig. 361c). It pushes against plank 'E', which is linked to the slide and partly covers the chamber floor as the sluice is not large enough to carry it all away. This reduces the load on the slide bars as they are covered by water, removing friction which prevents the slide from moving. Water now pushes it shut, closing the opening. The water which is running towards the sluice is turbulent because the sluice is closed suddenly and pushes up the lower gate. As it rises, more water gathers behind it and closes the gate.

The lock is now as shown in Fig. 360c. The boat which rested on the carriage floats free and is pulled into the upper canal and replaced by another.

When this has happened, the attendant can again set the water wheel 'D' in motion but this time he opens the sluice controlling the other side of the water wheel. The shaft 'C' is turned in the opposite direction, raising the gate. At first this causes no change to the movable parts or to the water level. However, when the top of the gate has almost reached the level of the upper canal, then the shaft 'C' pulls the chain 'F' such that the slide bars can move along the floor of the chamber, opening the sluice and allowing it to empty. This allows the lower gate to fall down, connecting the inclined track in the chamber to the inclined plane. The carriage, on which the boat now rests, can then be lowered.

This is a description of a typical operation with a quick turn-around of boats and with the minimum staff. Anything different would result in greater loss of water and Chevalier suggests that there could be the possibility of damage to the plant.

Something needs to be said about the motion of the carriages on the incline. When a loaded boat descends, it must have sufficient weight to pull up the other carriage and an empty boat. However, this is only the case on the twelve inclined planes on the eastern end of the canal. On the eleven western inclines, loaded boats have to be drawn up from the valley of the Delaware and empty ones lowered. Here it is necessary to use additional power to raise the loaded boats. Then the large high-breast water wheel 'B' which turns the drive wheel under the lock chambers is used to power the right or left incline depending upon the engagement of the [sliding] gears fitted to the common drive shaft.

Near the end of the descent, the carriage and the boat resting on it enter the lower canal, diminishing the counter-balancing effect. To counteract this, the slope of the track within the

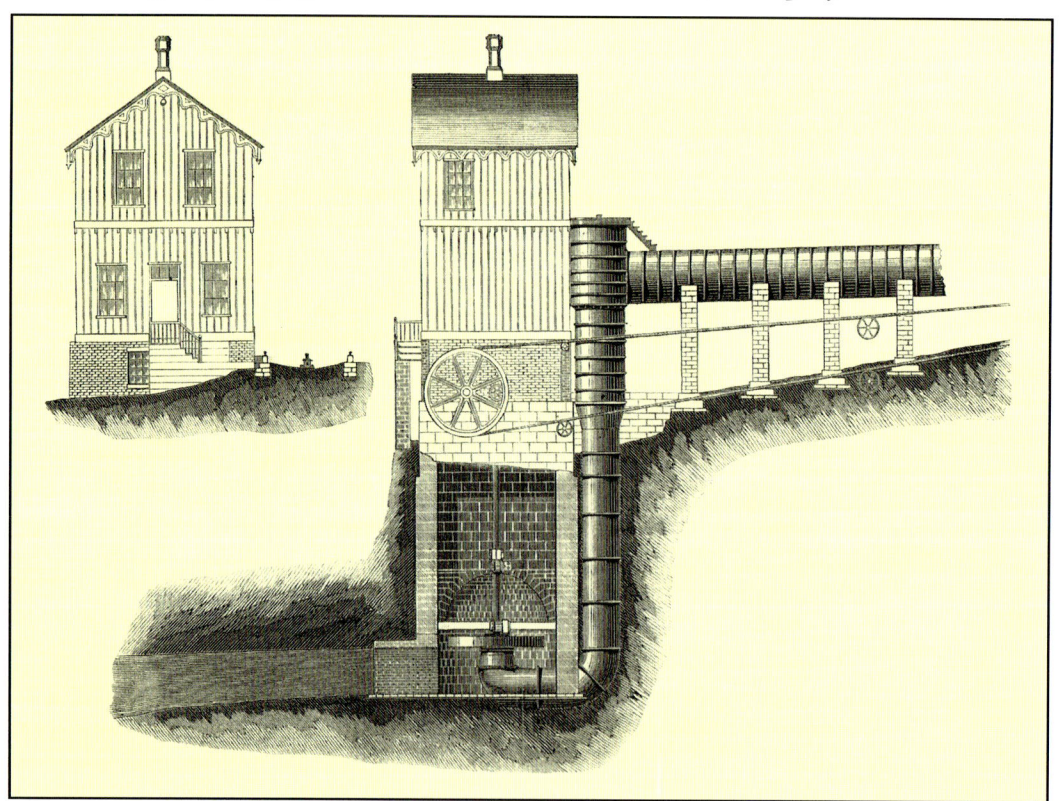

Fig. 55 Details of a turbine and turbine house constructed during the later development of the Morris Canal.

lock chamber is less than on the incline. Thus it needs less effort to draw the ascending carriage to the end of the track. The water wheel is used for this. There is a different problem for the descending carriage. From the description of the system so far, it cannot draw the descending carriage a sufficient depth into the water so that the boat resting on it would float off. Because of the total weight of the carriage and the boat, it would not be easy to provide sufficient power, such as by using a horse, to bring the carriage to rest at the end of the track. To overcome this, a system is used which was suggested by Fulton in 1796. Both carriages are connected by a second rope or a chain which passes over a wheel at the foot of the inclined plane. By means of this chain, the ascending carriage pulls the descending one more deeply into the water when the weight of the latter is reduced on entering the water.

Chevalier says that this second or bottom chain is considerably weaker and only forged from bar half the size of the main chain. It was not fitted to the centre of the carriage, but rather to the side closest to the other track. In Fig. 361 this chain and the wheel around which it is wound can be seen as well as the frame which carries the latter. Fixing the chain to the side of the carriage ensures that the wheel and its frame is not in the middle of the track, as this would cause problems in getting the boats on and off the carriage.

Chevalier says that, including fastening and releasing the boat from the carriage, on an incline

Fig. 56 The Morris Canal incline at Newark towards the end of the nineteenth century.

with a fall of 77 feet, 97 such operations have taken place per day, and that more boats could have been transported if they had been waiting.

The boats prior to 1835 were about 70 feet (21.34 metres) long, 9 feet (2.74 metres) broad and could carry about 25-30 tons.[40]

The first boat sailed between Newark and Easton on the 4th November 1831. Initially, not all the inclines were fitted with locks at their upper end and, in 1835/36, all those with dry summits were rebuilt with lock chambers so that a larger size of boats could be used.[41] In 1845, sectional boats began to be used. These

Plane Car, with boat at top of Plane, near Phillipsburg, N. J.

Fig. 57 A postcard of a boat at the top of the Philipsburg inclined plane. Note the water supply trough serving the turbine in the background.

were 87¹/₂ feet (26.67 metres) long, 10¹/₂ feet (3.2 metres) wide and carried up to 70 tons on a draught of 4¹/₂ feet (1.37 metres). Because they could be divided at the inclines and locks, a reconstruction of all the inclined planes was begun on a simpler design using a dry summit. The water wheels were also replaced by more efficient turbines, which drove the winding drums through a gear train. The reconstruction was completed by 1860.[42]

Since the Morris Canal, almost from the start of its operation, was in direct competition with the railroad, it is amazing that even after its last major modernisation it still made a profit, and in 1866 it could carry a record 889,220 tons of goods.[43] From 1875, when the Lehigh Valley Railroad opened, there was a rapid decline in traffic using the canal and just a year after nationalisation in 1923 it was closed. The canal decayed but some remnants of the inclines have survived. A water turbine has been set up in a small park on Lake Hopatcong, formerly the main reservoir for the canal, as a monument to the old system.

Since 1989 there have been efforts to restore to its original condition the well preserved 80 feet (24.4 m) rise Incline 4 West at Waterloo near Stanhope, and to develop it for tourism. It is being organised by a local museum, which interprets New Jersey's history in the eighteenth and nineteenth centuries.

The inclines on the Morris Canal are most important because they were used as the model for later ones in East Prussia (now Poland) and Japan.

The Pennsylvania Portage Railroads

Inclined planes associated with waterways were built elsewhere in America, in particular on the Pennsylvanian waterway system which lay to the west of Philadelphia. A waterway route to the Ohio was sought but the hills around Philadelphia and the Allegheny mountains formed a formidable barrier. Two railways, using inclined planes because of the lack of power of contemporary steam locomotives, were built to connect existing waterways.

The Columbia & Philadelphia Railroad connecting the Susquehanna and Delaware valleys opened in April 1834. It was 82 miles long and had an incline at either end. Belmont incline at the eastern end was 2,805 feet (855 metres) long with a rise of 187 feet (57 metres). At the other end, the Columbia incline was 1,800 feet (549 metres) long with a 90 feet (27.4 metres) rise.

In competition was the Allegheny Portage Railway linking Holidaysburg on the Juniata with Johnstown on the Connemaugh. It was 37 miles in length and had ten inclined planes. The five eastern ones overcame a rise of 1,399 feet (426 metres), those in the west 1172 feet (357 metres). The railway opened in March 1834.

Some six months after the railways opened, a boat was carried over the Portage railway and this led to the design of sectional boats. These could be placed on trucks for movement over the railway and assembled once they were returned to the water over slipways which were provided at either end of the railways. Captain H. A. Walters described the boats as follows:

Fig. 58 One of the locos used on the Portage Railroad.

The sectional boats were 82 feet in length, 13 feet in width and in depth 12 feet and were divided into four sections each 20$\frac{1}{2}$ feet long. The boats were round in their bottom. The sections were fastened together by irons about halfway down the side — the iron projected out from one section into a V-shaped iron on the other section, then a T-iron fitted down to both of these irons and locked them together. One section was placed upon one railroad truck which was a bit longer than the section — about 23 to 24 feet — and had four wheels. The trucks were round in the bottom to fit the boat sections.

The width of the boat seems wide for use on a railway but perhaps there was a one-way system of operation. Charles Dickens travelled over the railway [by boat].

There are 10 inclined planes, 5 ascending and 5 descending; the carriages are dragged up the former and let slowly down the latter by means of stationary engines; the comparatively level spaces between being traversed sometimes by horse and sometimes by engine power as the case demands.[44]

Henry S. Tanner's book Canals and Rail-roads of the United States, published in 1840, describes the inclines.

…and on the inclined planes, and along the canal basins, at the two terminations of the road, flat rails upon timber are used. At the head of each inclined plane, there are two stationary steam engines of about thirty-five horse power each, which give motion to the endless rope, to which the cars are attached. Only one engine is used at a time, but two are provided to prevent delay from accidents. Four cars, each loaded with 7,000lb, can be drawn up, and four may be let down at the same time; and from six to ten such trips can be made in an hour. A safety car is coupled to the cars, both ascending and descending and stops them in case of accident to the rope, which adds greatly to the security.

The inclines varied in length from 1,500 to 3,100 feet (457 to 945 metres), the slope varying between 1:16 and 1:10, and the highest point was 2,334 feet (711 metres) above sea level. The system was abandoned on 1st July 1855.

Today, part of the line is a National Park, a house at the top of No.6 Plane being used as a Portage Railroad Museum.

The Grand Western Canal incline

Meanwhile, the building of small tub-boat inclines continued in England. In 1836 on the Grand Western Canal, which we will visit again on page 116 to look at its numerous vertical lifts, an incline came into service constructed by James Green at Wellisford with a slope of 1:5$\frac{1}{2}$ and a rise of 81 feet (24.7 metres). It did not enter service immediately as it did not work properly. Green had designed and built it according to Fulton's 'bucket' system and the buckets and their contents, some ten tons, were too light. Green was dismissed and a steam-engine had to be installed by W. A. Provis before the incline could begin work after a two year delay. The tub-boat section of the Grand

Fig. 59 A recent view looking up the line of the incline at Ilminster on the Chard Canal.

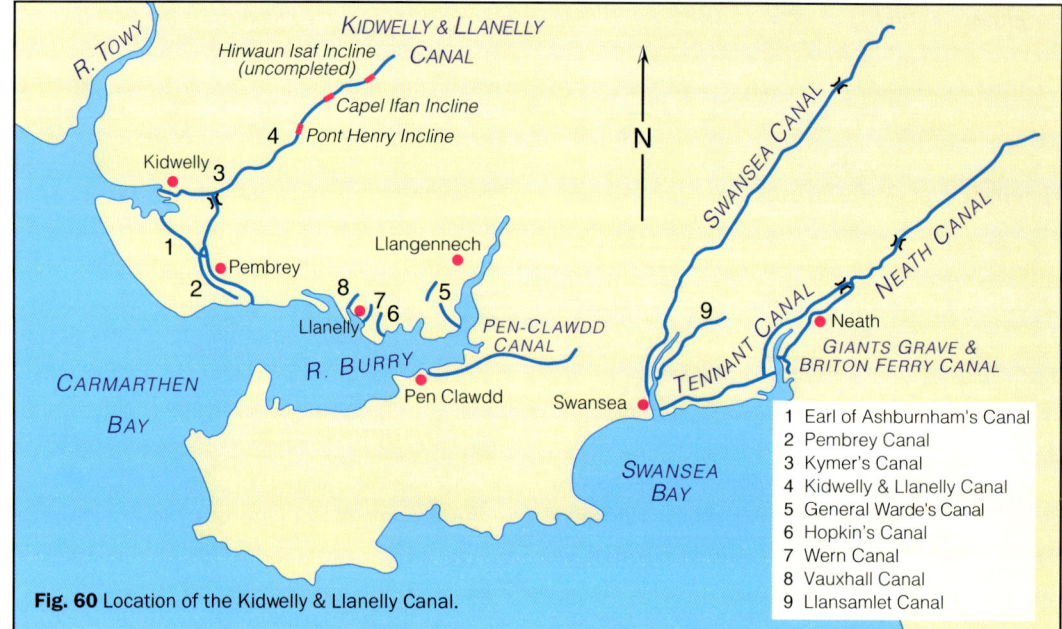

Fig. 60 Location of the Kidwelly & Llanelly Canal.

Map labels:
- R. TOWY
- KIDWELLY & LLANELLY CANAL
- Hirwaun Isaf Incline (uncompleted)
- Capel Ifan Incline
- Pont Henry Incline
- Kidwelly 3
- 4
- Llangennech
- Pembrey
- 1
- 2
- 8 7 5
- 6
- Llanelly
- PEN-CLAWDD CANAL
- R. BURRY
- Pen Clawdd
- Swansea
- CARMARTHEN BAY
- SWANSEA CANAL
- N
- NEATH CANAL
- TENNANT CANAL
- 9
- Neath
- GIANTS GRAVE & BRITON FERRY CANAL
- SWANSEA BAY

1 Earl of Ashburnham's Canal
2 Pembrey Canal
3 Kymer's Canal
4 Kidwelly & Llanelly Canal
5 General Warde's Canal
6 Hopkin's Canal
7 Wern Canal
8 Vauxhall Canal
9 Llansamlet Canal

Western Canal was closed in 1867, the lifts ceasing to operate at this time.

The Chard Canal

On this 13$\frac{1}{2}$ mile canal in Somerset, which runs southward from the Bridgwater & Taunton Canal, four inclined planes at Thornfalcon, Wrantage, Ilminster and Chard Common came into service in 1841/42. Only on the uppermost at Chard Common, rise 86 feet (26.2 metre) and single-track, were boats carried on a carriage. On the remaining three inclines water-filled caissons were used in which the boats 'swam over the summit'. This was the earliest use of caissons on inclined planes and is described later on page 69. The winding gear on the inclined plane at Chard Common was powered by a turbine, which had a head of 25 feet (7.6 metres) and used 725 ft³/min of water. Though the inclines and canal were closed in 1868, the tunnel at Wrantage and the slope of the inclines at Wrantage and Ilminster can still be seen.

The Kidwelly & Llanelly Canal

On this Welsh canal, three inclined planes designed by James Green should have come into service in 1838. Only the lowest two inclines at Pont Henry, 57 feet (17.4 metre) rise and Capel Ifan, 56 feet (17.2 metre) rise, were opened.

They were probably double tracked. The third incline, at Hirwaun-isaf, was to have an 84 feet (25.6 metre) rise. On this nine mile long canal, tub-boats were used which carried up to 6 tons. Traffic had ceased by 1867, but besides this, little is known about the inclines.

The Elblanski Canal

On the 29th October, 1860, in East Prussia, the Elbing Oberland Canal (today known as the Kanal Elblanski) came into operation using four inclined planes. Today they are the world's oldest working inclines where the boats are carried out of water.

A waterway from the Oberland to Elbing was first suggested by local land owners in 1825 in the province of East Prussia. The numerous Oberland lakes had been linked by canals with four locks (dimensions 31.4 by 3.14 metres) in 1844-1850. The water levels of the lakes had been considerably lowered, Sammroth Lake and Pinnau Lake by over 5 metres and the reclaimed land used for meadows and pasture. The ascent from Lake Drausen, near Elblag (Elbing), to the 100 metre higher Oberland was initially by 5 locks (total rise 13.5 metres). For the remaining 85.6 metre rise in just 7.2 km the waterway administration had proposed 15 locks, including four double, one triple, one six-fold and one seven-fold staircase sets of locks. In all 32 lock chambers were required.[45] Instead, four inclines, based on those on the Morris Canal, replaced the 15 locks and staircases. The engineer was Steenke, who had studied their construction and operation during a visit to America in 1850. The design of his inclines is enchanting and their magic still casts a spell over people, and not just industrial historians. Germany's most important hydraulic engineer of the nineteenth century, Gotthilf Hagen, included in his work on the subject[46] a complete chapter on the Oberland Canal from which the following passages are taken.

When Baurath [design engineer] Steenke received an order in 1837 to draft a scheme for this canal, he was certain that such a difference in levels on such a short canal could not be overcome by means of ordinary locks.

As an appropriate method for overcoming these difficulties, he decided to use a system based on the Morton's patent slip where the boats are put on carriages and are pulled up on railway tracks. If it was possible in this way to lift a completely fitted-out East-Indiaman, then there could be no doubt that a loaded canal boat could be treated in the same way. With this

thought inclined planes were suggested, which Steenke was soon afterwards to learn about on his study trip to America.

The canal originally had four inclined planes, while the first 45 feet (13.72 metres) of rise from Elblag was by five wooden locks. They were later replaced by a fifth incline. The rise of the inclines, listed with Polish and German names from the summit, is: Buczyniec or Buchwalde 67 feet (20.42 metres), Katy or Kanten 62 feet (18.90 metres), Olesnica or Schönfeld 80 feet (24.39 metres), Jelenie or Hirschfeld 72 feet (21.95 metres) and Calony Nowe or Neu-Kussfeld 45 feet (13.72 metres).

The canal system is not connected directly to any other waterway, so the boats could be designed to suit the inclines. With the exception of some small steam and sailing boats, they were of a standard size (see Fig. 71). Regulations limited their dimensions to 80 feet (24.48 metres) length, 9.78 feet (2.98 metres) width and a maximum draught of 3.6 feet (1.10 metres). They carried around 50 tons, though loads sometimes reached 70 tons on a draught slightly more than specified. On the lakes which formed part of the system, boats used sprit, gaff and occasionally lateen sails, though they were hauled by men on the canal. The boats also had

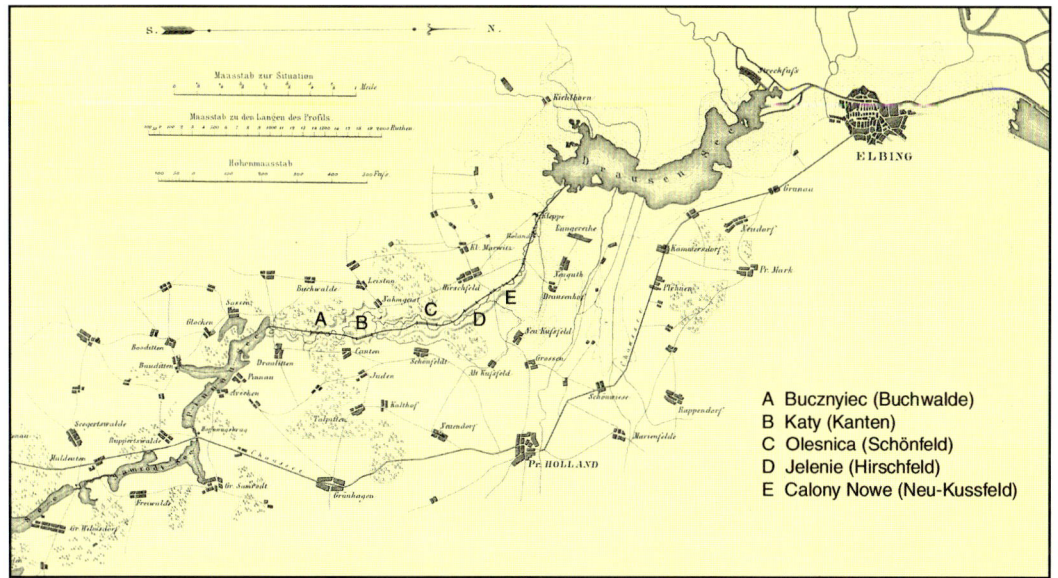

Fig. 61 A contemporary map of the Elbing Oberland Canal.

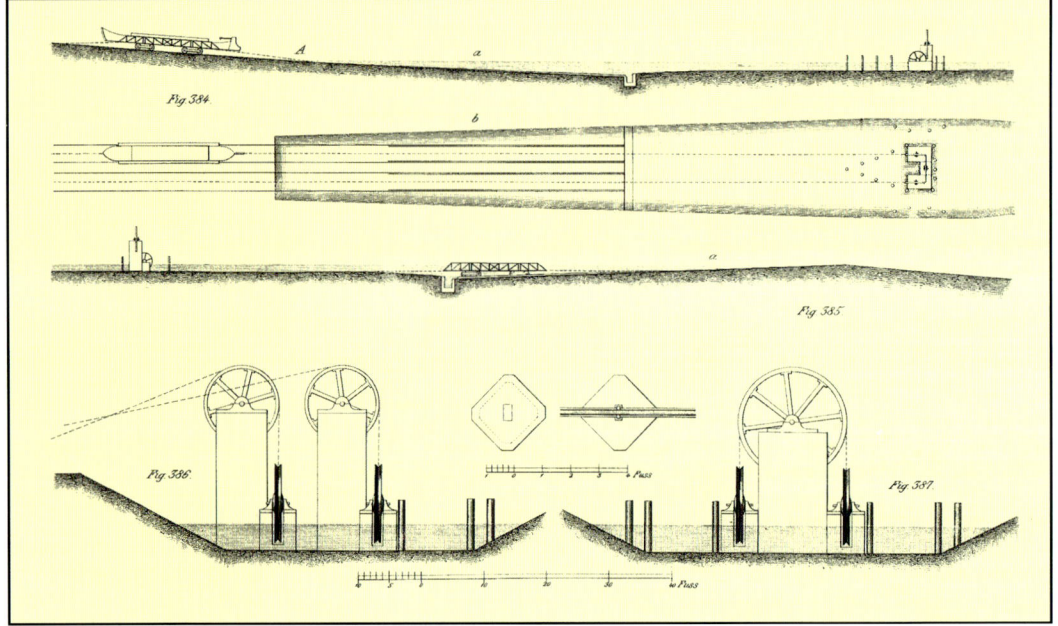

Fig. 62 Cross-section and plan (fig. 384) of the lower end of an incline on the Elbing Oberland Canal, with a cross-section (fig.386) of the upper end. Below are cross-sections of the upper (fig.386) and lower (fig.387) rope guide pulley system.

a rowing boat and this hung from an iron crane at the side of the carriage on the inclines (see Fig. 74). A skipper, mate and a boat's boy formed the crew, which was often just a family.

The inclines are all of basically the same design, with two parallel tracks of gauge 10 feet 9 inches (3.27 metres). The carriages are also similar, except for those at Buczyniec where the side platforms (see Fig. 64) are higher to allow for a possible variation of 5 feet (1.60 metres) in the summit water level.

The slope of the incline from the upper canal to the summit and the bottom section of the lower incline is 1:24 (see Figs. 62 and 70). After a carriage has passed over the summit and starts to descend, its weight helps to raise the ascending one so the slope of the incline can be increased to 1:12. Where the level of the

upper canal remains constant, the summit needs only be around one foot (0.31 metres) above water level. At Buczyniec this is greater to allow for variation in the summit water level.

The boat must, while it is being raised from the water and lowered therein again, be held horizontal or at the same angle at which it floats, so that neither end is immersed more deeply than usual and thus exposed to any especially high pressure. For this [reason]… a second set of rails are placed alongside the existing track and a second wheel rim, fitted to the wheels which run on the main track, engages with this second short track. In this case, the requirement is met as shown in Fig. 390a [see Fig. 65], in that all wheels fitted to the carriage are provided with two separate wheel tyres on either side of a flange.

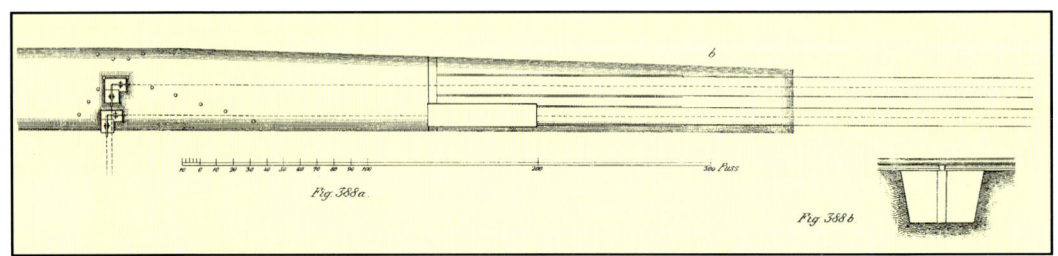

Fig. 63 Plan of the upper end of an incline (fig.388a) and a cross-section through one of the concrete track supports (fig388b).

The inner tire of the wheels at one end of the carriage run on the main rails, while at the other end of the carriage it is the outer tire which runs on them, the wheels at one end being slightly closer together than at the other. As the second rail is on the inside of the track on the lower incline and on the outside of the upper one, then either one end of the carriage or the other will be raised to keep it level, depending upon which part of the incline it is on. The carriage sits on bogies 30 feet (9.10 m) apart and to keep the carriage horizontal, the secondary tracks have been laid with their first 30 feet horizontal and then angled with the same slope as the main track, though somewhat higher.

Power is provided by water wheels supplied with water from the upper canal. They have a diameter of 28 feet (8.47 metres), a width of 12 feet (3.77 metres) and drive 12 feet (3.77 metre) diameter iron drums which are 7 to 8 $\frac{1}{2}$ feet (2.20-2.67 m) wide depending upon the length of the incline. To change the direction of the drive, the main gear is taken out of mesh and an intermediate idler gear engaged. Two wire ropes, their ends fastened to the carriages, are attached to and wound in opposite directions around this drum. Thus one rope pulls one carriage upwards, while the other unwinds and allows the other carriage to descend. The drum has a spiral grove so that the ropes move from one side to the other as they wind on and off. To ensure that the carriages reach the end of the tracks, they are connected at their lower end by a somewhat weaker third wire. The water wheel pulls the ascending carriage and by means of the third wire hauls the other carriage out of the upper canal to the summit. (see Figs. 62, 63 and 68 for the layout of the ropes)

The power can be increased or decreased by lifting or lowering a sluice controlling the water. The greatest power is required at the start of the operation and this can be reduced when the carriage coming from the upper canal has crossed the summit, the amount of reduction depending on whether the boats are loaded or empty. If a fully loaded boat descends and an empty one or just the carriage ascends, then there is a strong counter-balance effect; little power is necessary and the operator must apply the brake fitted to the gear shaft.

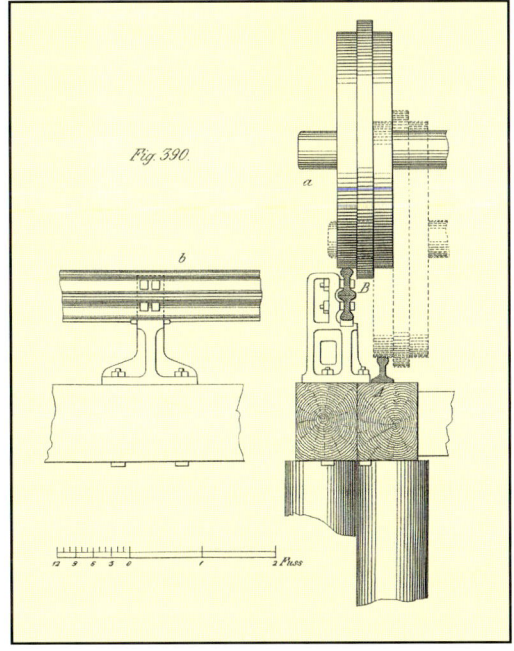

Fig. 64 & 65 The wheels on the carriage have a central flange. Those at one end of the carriage are also slightly narrower in gauge, as seen in the elevation and cross-section of the self-levelling system of rails at top and bottom of the incline. The self-levelling rails are on the inside at one end of the incline and on the outside at the other.

Fig. 66 A boat descending an incline passes the ascending empty carriage. Below, a second boat is manoeuvering around the return pulleys so that it is ready to enter the carriage once the descending boat has floated clear.

The machine house stands to one side of the upper canal and from here both ropes are led to the 12 feet (3.77 m) diameter iron wheels, which are shown in front view in Fig. 386. They are, Fig. 385, placed slightly out of line so that the ropes do not touch. Following the line of the grooves, both ropes go vertically downward some 14 feet (4.39 m) to two other similarly-sized wheels which are situated at right-angles to the first. These are shown from the side, and they guide the ropes to close to the bottom of the canal and of course parallel to its direction and the inclined plane.

Fig. 385 shows the upper end the incline plane with part of the upper canal and the summit. A carriage is also shown which is deep enough in the water for a boat to be floated onto it. Notice also that the masonry which carries the wheels blocks a large part the canal and this has to be widened here to leave enough space at one side for the free passage of boats.

On its passage over the inclined plane, each wire is carried by light iron rollers every 30 feet (9.42 m) to prevent it from rubbing on the ground and quickly wearing away.

Fig. 384 again shows the cross-section and plan of the lower end of the inclined plane with a section of lower canal. Note that at 'A' the slope changes from 1:24 to 1:12. The carriage loaded with a boat travelling upwards now has this inclination up to the summit.

The rope is guided into the lower canal. Both ropes lie here just as far apart as the centre lines of the two tracks, namely 17 feet (5.34 m) and are led at an equal distance from the canal bed into the grooves of two vertical 12 feet (3.77 m) diameter wheels around which they are wound in a vertical direction to pass over a third, 17 feet diameter wheel shown in Fig. 387. The masonry, which carries these three wheels, is in the middle of the canal and boats can pass round it on either side as the canal is widened here appropriately. However, so that the boats passing them do not push against the walls, just as on the upper canal, planks are placed here against which the boatmen may push with their boathooks.

The result of this is that the whole rope system from the drum, over the inclined plane, and back to the drum is completely closed. However, it must always be kept at the correct tension so that when a carriage goes over the summit it is not allowed to run downwards and is then suddenly held back by the rope, causing it to break. To prevent this, the ropes are connected to the carriage by strong wedged hold-fasts.

The middle of the inclined plane is marked exactly. Both carriages are lined up here, and where the tension is to be regulated, both wedges are opened and the ends of both ropes pulled together as tightly as possible by a pulley block, after which the wedges are closed. If new ropes, which have previously been tested to three times their highest possible tension, are installed it is at first necessary to tighten them repeatedly and then it is rarely necessary later.

The ropes were made by Felten et Guillaume of Cologne. The main ones have seven strands, each of seven Nr. 9 wires. The lay or length of twist must not be too short (it was 0.35 to 0.37 metres), because otherwise the wire would stretch too much.

The rails were the same as those used by the Ostbahn (Eastern Railway). On the main track they were originally spiked to longitudinal oak sleepers with cross-ties every $3^1/_2$ feet (1.10 metres). The wooden sleepers, cross-ties and rails needed constant repair and derailments were not unusual because of gauge widening.

Steenke tried placing the sleepers on flat-topped four-sided concrete pyramids. (see Fig. 63)… To fasten the rails to them, dowels of pine are put in the mould in which the concrete blocks are formed. These dowels will be held tightly after the block has hardened, and the spikes can then be fastened securely. The blocks are arranged diagonally …

The carriages weigh about 500 Centner (25 tons), and on the Buchwalde incline about 520 Centner (26 tons), an empty boat weighs 160 Centner (8 tons) and its cargo can be up to 1400 Centner (70 tons). The gross load of a carriage can therefore be 2,000 or even 2,400 Centner (100-120 tons). Each of the eight wheels therefore carries 275 Centner (14 tons). At first their iron tyres failed after a few weeks and had to be replaced immediately by steel, which have barely the necessary hardness and become worn out faster than any other part of the plant.

The rails needed constant replacement and for this reason, cast steel rails were used later.

Underwater, the sleepers were fixed on piles. The side rails were on cast-iron chairs, shown in Fig. 65. The chairs have four different heights, so that the height of the rails increases progressively. To strengthened them and to prevent distortion,

Fig. 67 A sailing-boat on a carriage passing over the summit of an incline around 1900.

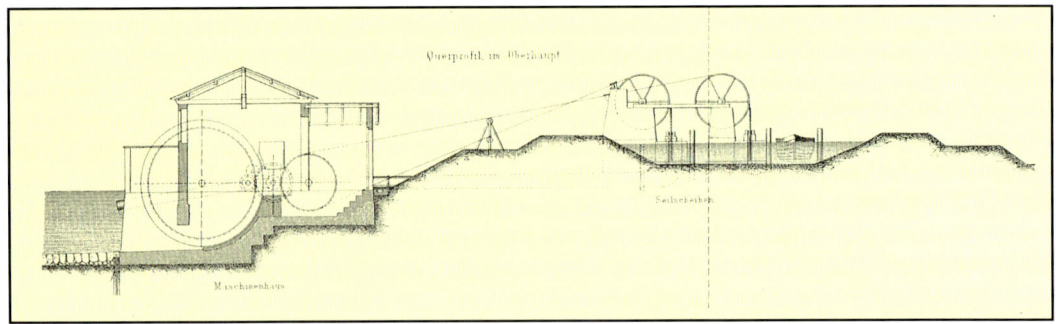

Fig. 68 Cross-section through a machine house showing the water wheel and the drive to the incline.

a second rail of the same form was riveted beneath.

The rails at the end of the inclines were bent upwards so that the carriages did not run off the end of the track. However, the machine operator could see the position of the carriages from marks on the drum and rope and can stop them as soon as the boats can float off.

The carriages are made from iron except for the wooden planks on the floor on which the boat sits and the walkway. The floor is not completely flat, but is slightly curved in the longitudinal direction, matching the shape which the boats assume when settling onto the carriage.

The carriages sit on two four-wheeled bogies which can pivot to stop the wheels from leaving the rails when passing over a summit.

The front and rear wheels in each frame are carried by iron girders on edge into which their axles are fitted. The iron girders on either side were not at first connected, which resulted in the problem that sometimes they did not remain parallel and this caused derailments. Recently, cross bars have been fitted through the ends of both girders and held by screws.

The brakes, which work on all eight wheels, are controlled from footways on the sides of the carriage. The iron derricks for rowing boats are also fitted here.

The whole arrangement was designed and executed by Director Krüger of the Dirschau (Tczew) Engineering Institute and since the opening of the system has proven satisfactory on all four inclines. It has been found that the operator, without changing his position and

without outside help, can raise the boats into the upper canal or lower them into the bottom canal.

As soon as the bell gives the signal that both boats have been pulled over the carriages and are fastened to them and having engaged the correct gearing for turning the drum in the right direction, then the operator lowers the sluice by means of the crank and reduces or shuts off the influx as soon as one carriage has passed the summit. However, he grasps the brake so that, if the speed increases too much, he can control

Fig. 69 The guide pulleys at the top of an incline, with the waterwheel and machinery house behind.

Fig. 70 A view from the bottom of an incline. The lesser angle on the lower section allows for the fact that the machinery must also raise a carriage out of the water at the upper end of the incline. Once this is over the summit, its weight will assist the machinery in raising the second carriage, so the angle can become greater. The bridge on the right is over the channel which feeds water used by the waterwheel back into the canal system.

the movement of the machine very surely. From the number of winds of the ropes on the drum he can also see the position of both carriages, and thus he is in a position to know how far to let them go so that the boats float clear. However for the last part of the way, particularly when a boat, which is more heavily loaded than the other, is almost floating, a little further help is given by the water wheel. . . .

As soon as a carriage and boat are set in motion, the boat's mate must climb onto the walkway and during the whole trip over the inclined plane must remain standing beside the brake, so that this can immediately be applied should the rope break. The brake is certainly not so strong that it would bring the carriage with loaded boat to a standstill on the steepest part of the incline, but it can decrease the speed enough so that any danger disappears. Up to now, it is recorded that such an accident has not taken place.

The time for the passage of a boat over an incline, including loading and unloading of the carriages, is about a quarter of an hour. A boat needs therefore approximately one hour to pass over the four inclined planes, whose descent totals 273 feet (85.68 m). To overcome this in the ordinary way through locks, would have

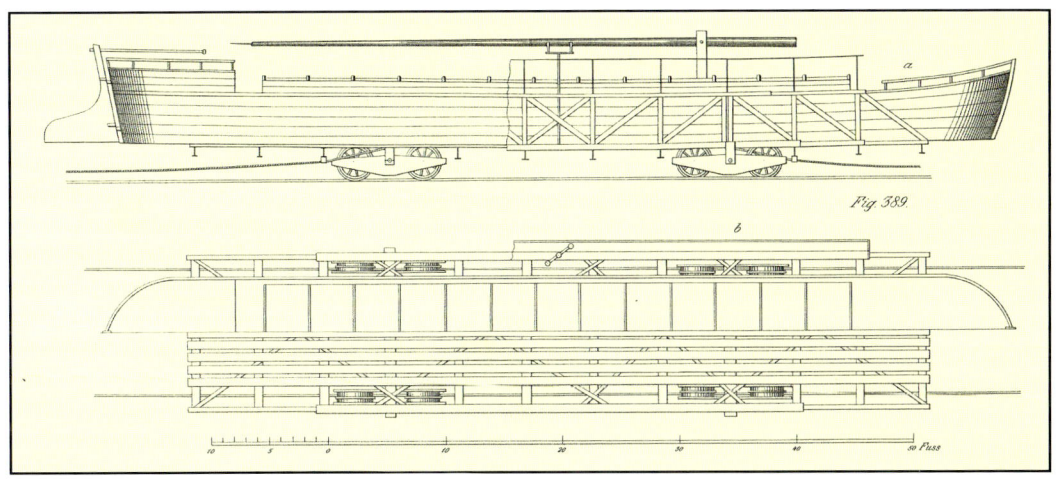

Fig. 71 Plan and elevation of a carriage and boat.

needed at least twenty-three locks and if the passage of each took a quarter of an hour, so it would have required at least six hours for their ascent or descent. This saving in time is one advantage of inclines over locks. On the incline at Buchwalde there are already some 73 boats passing every day, a total, in both directions, of 146 vessels transported, which is many more than chamber locks could pass.

A second benefit is the saving of water, and here again, up to now, there is no problem even if the water levels of the upper section, because of a long persistent drought, were somewhat lower than had been expected. If this disruption should happen often, it would be possible to completely overcome the lack of a supply of water by the application of steam power. At most times, and particularly in the summer and autumn, there is an ample supply available, so that on average about 10 cubic feet per second (0.31 m³/s) could also be used for industrial purposes.[47]

There is little to add to Hagen's description; that the water wheels produced 68 HP (50 kilowatt) and consumed about 1 m³/s water. The water overflow into the lower canal is through channels, which are provided with cascades. No detrimental effects have been

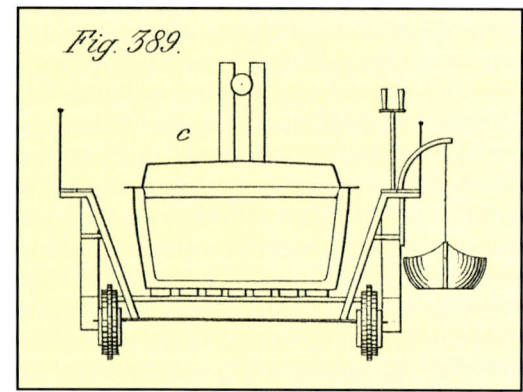

Fig. 72 Cross-section of a carriage and boat.

observed on the boats due to transport 'in the dry' on these inclined planes.[48]

When opened, the Elblanski Canal had a length of 116 km, of which only approximately a quarter was canal. Twenty-one years after it opened, the five wooden locks at the lower end of the waterway were replaced by a fifth inclined plane at Neu-Kussfeld (Calony Nowe; rise 13.5 metres). In contrast to the other inclines, it was powered by a turbine.[49]

Today you can experience the inclines on the Elblanski Canal by taking a day trip on board one of the passenger boats which ply regularly between Elblag and Ostróda in the summer months.

Fig. 73 Drive sheaves at the top of Buchenwald inclined plane around 1900.

If you do not have the time for this excursion, the uppermost inclined plane at Buczyniec (Buchwalde) is easily reached by car. Driving from Ostróda to Elblag, you turn off main road 7 about 30 km north-west of Ostróda in Marzewo (Mokrau)and take the Drulity (Draulitten) road. After eight kilometres and beyond the village you turn right onto a cobbled road signposted to Buczyniec. After one and a half kilometres turn off to the left onto an unpaved road at whose end, about 200 metres away, lies the inclined plane. The operator will show you the power plant and explain its operation. Near the power house you can find the commemorative stone erected in memory of Georg Steenke. During the summer,

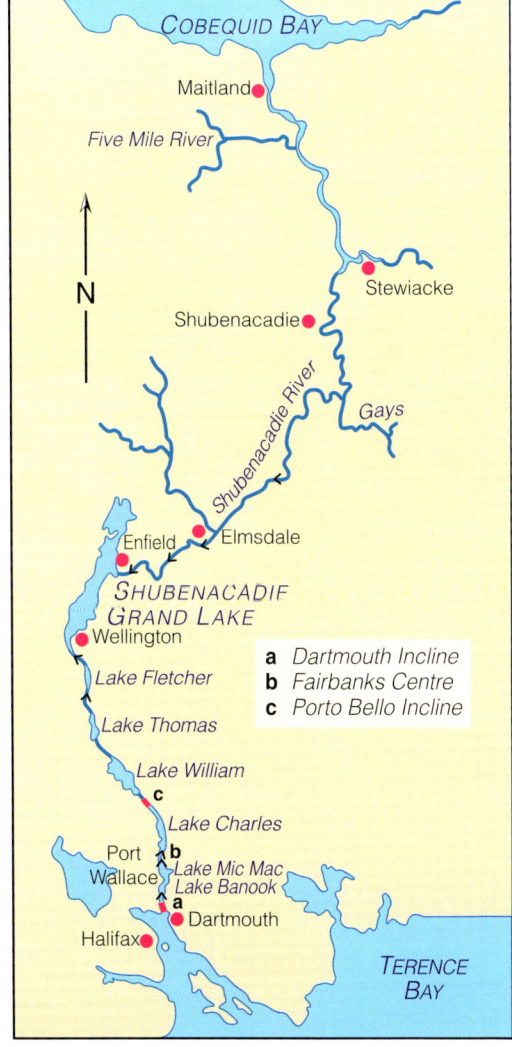

Fig. 74 Location of the Shubenacadie Canal.

the trip boat from Elblag passes Buczyniec (Buchwalde) at 12-30 and that from Ostróda at 14-30. These are the best times to see how a dry inclined plane operates.[50]

The Shubenacadie Canal

After several attempts between 1826 amd 1831, the Shubenacadie Canal in Canada was finally built between 1854 and 1861. This 86 km long waterway in Nova Scotia linked the Atlantic to the Bay of Fundy. Its route followed, to a great extent, lakes and rivers and was similar to the Scottish Caledonian Canal. There were nine locks and two inclined planes, Dartmouth, rise 60 feet (18.3 metres) and Porto Bello, rise 35 feet (10.7 metres). These were similar to those on the Morris Canal and had equivalent rises. The canal was designed for boats of 66 feet (20 metres) by 16.5 feet (5 metres) by 4 feet (1.22 metres) draught. However, after nine years the canal was closed because it was uneconomic. Today the best display about the history the canal is at the Fairbanks Centre.[51]

The Canal de l'Ourcq

In France, about 35 km to the east of Paris the 108 km long Canal de l'Ourcq runs close to the Marne which is at a lower level. The canal is 12.2 metres higher and, near the city of Meaux, only 550 metres from the river but, to get from the canal to the river, boats had to make a 100 km long trip via Paris. So in 1884 a dry incline with a summit was built with a single track (1.94 metre gauge) sloping at 1:25. The boats to be transported (length 28.0 metres, width 3.1 metres and depth 1.2 metres at a load of 70-75 tons) were carried on a 24 metre long carriage. This had two four-wheel bogies each with a wheelbase of 2.0 metres. The basic operating method was similar to that on the Elblanski Canal inclines.

Between 16.5 and 25 HP was needed and this was provided by a turbine supplied with water from a 1.6 metre high weir on the Marne at Basses-Fermes. Originally the drive was by a chain and sprocket wheel, but this was not

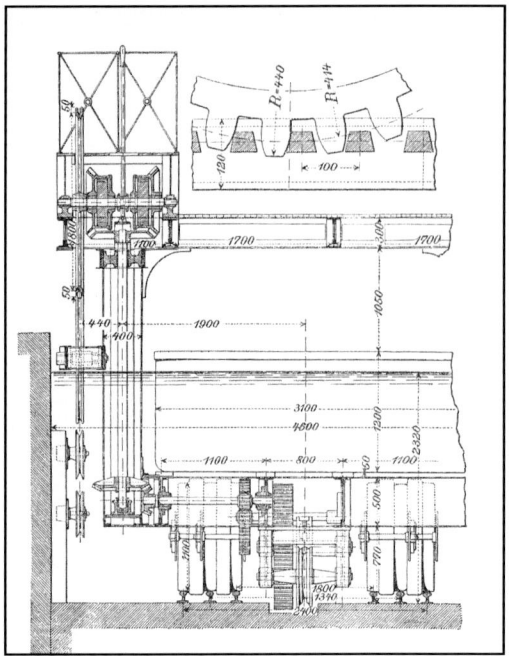

Fig. 75 Cross-section through the drive mechanism of the inclined plane at Meaux.

successful and it was quickly replaced by a Riggenbach-type geared track which proved very reliable. To control the operation, only one man was needed and everything concerning the movement of the boat was undertaken by its crew.[52] The incline was closed in 1922[53] and today nothing survives. The long trip between canal and river remains and recently plans have been produced for the construction of a pleasure boat lift about 25 km above Meaux at Lizy-sur-Ourcq.[54]

Japanese Inclines

A few years after the inclined plane at Meaux came into service, at the other end of the world in Japan, two inclines with dry operation were built. Their designer, Sakuro Tanabe (1861-1944), had, just like Steenke, made a two-month-long study trip to the USA in 1888/89 to see technological developments there. The most important result of this journey was the

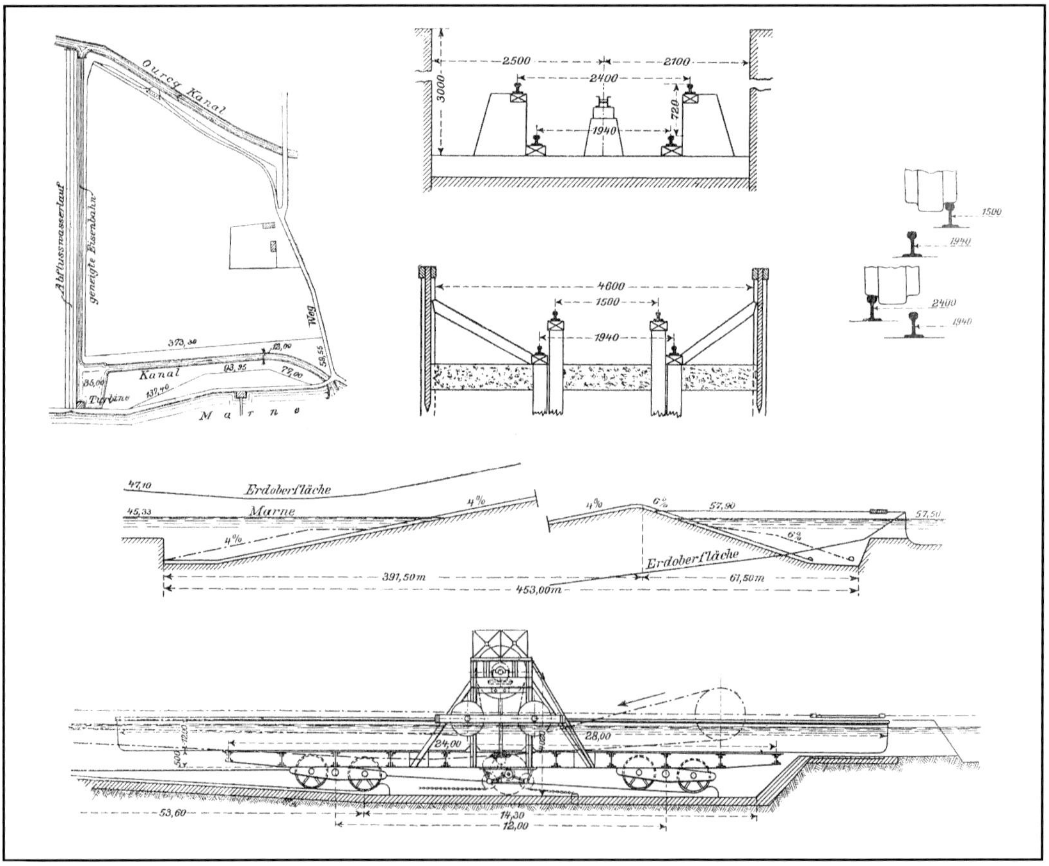

Fig. 76 Location and cross-section of the inclined plane at Meaux.

Biwako Canal which included an incline and hydro-electric power plant which began work at the same time as the earliest American plant. Tanabe proved himself to be brilliant in other engineering disciplines and he was presented with the Telford Medal of Britain's Institution of Civil Engineers.

Japan's largest inland lake, the 65 km long and 20 km broad Biwa-Ko, drains into Osaka Bay via the Yodo River. It was however only navigable from its mouth on the Bay to Fushimi, a distance of about 37 km. To the north of Fushimi lay the Emperor's city of Kyoto, which was poorly linked to the sea by the old Takasegawa Canal. With the building of the Biwako Canal, an 11.3 km long waterway was created, which had two locks, three tunnels, and the Keage Incline which overcame 118 feet (36 metres) of the canal's total rise of 140 feet (43 metres). The incline used twin operation, the gauge of the tracks being 8 feet (2.44 metres) and for the first time, in 1892, electrical energy generated nearby was used to power an incline. Completed in 1890, the canal took five years to build.

In 1894, the Takasegawa Canal was replaced by the more effective Kamogawa Canal, which

Fig. 77 Location of the Biwako Canal.

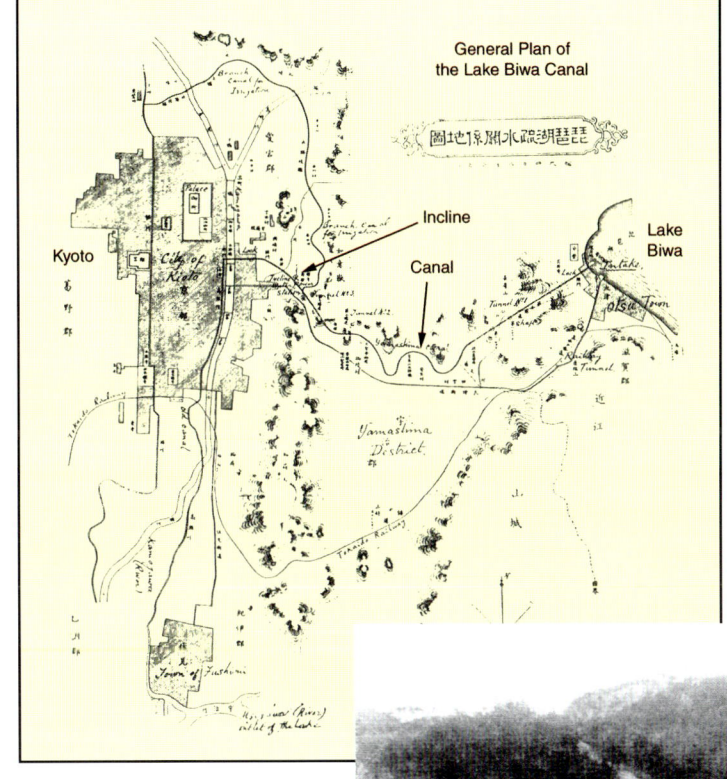

Fig. 78 The upper end of the Keage Incline, with the tunnel beyond, during construction.

Fig. 79 The Keage Incline around 1890.

ran parallel to the similarly-named river. This canal had, as well as eight locks, the second inclined plane in Japan, the Fushimi Incline, which had a rise of some 30 metres. Now, at last, Kyoto had a good inland waterway. The boats which used it were about 147.5 feet (45 metres) long, 7.5 feet (2.3 metres) broad and had a draught of 4 feet (1.22 metres).[55]

Traffic on both canals ended in 1914. Of the inclines, the one at Kyoto, the Keage Incline, is still well preserved and could, without too much expense, be put back into service again for pleasure boats. Only remnants survive of the one at Fushimi following the construction of a road across it in 1968.[56]

Fig. 80 The incline at Stanley Ferry around 1900. It was used by compartment boats on the Aire & Calder Navigation.

The Tom Pudding incline

The year 1891 also saw the construction of an incline at Stanley Ferry, near Wakefield, for use by coal-carrying compartment boats called *Tom Puddings*, which were unloaded using a hydraulic lift (see page 87).[57] The compartment boat system was used on the Aire & Calder Navigation and was designed in 1862 by its Engineer, William Bartholomew. They were winched up the incline and out of the water on special twelve-wheeled railway wagons on which they were pulled by a steam locomotive to St. Johns Colliery about one mile distant. The use of the incline lessened the handling of the coal and thus reduced the chance of breakage, an important factor in the design of the system. A more conventional loading system, tipping from railway wagons, was also provided near to the incline. This took over all coal handling here in 1941 when the colliery built new coal screens which could not accommodate the Tom Puddings.

The Trent-Severn Waterway

Two further dry inclines entered service in 1920 on this Canadian waterway. Since it also has an important role relative to vertical lifts (see page 113), the following details about the waterway will be useful.

The Trent-Severn Waterway extends for 240 miles (387 km), from Trenton on the Bay of Quinte on Lake Ontario, westward to Port Severn on Georgian Bay on Lake Huron. The main purpose for building the waterway, which was first suggested in 1785, was the opening up of the huge forest to the west of Trenton and to transport wood to the east by a waterway exclusively within Canada.

Hydrologically, the area is characterised by the two rivers, Trent and Severn, the former draining into Lake Ontario and the latter into Lake Huron. Both rivers were only partially navigable and their course was interrupted by numerous rapids. Furthermore their hinterland has 17 large lakes in which there are 160 large islands. The canal's summit on Lake Balsam is 256.3 metres above sea level, 181.9 metres above the Bay of Quinte, and 80.1 metres above Georgian Bay.

Constructing the canal began at Bobcaygeon in 1835 with the building of a wooden lock on the waterway's eastern descent. It ended in July 1920 with the opening of Couchiching Lock. The same month the excursion steamer *Irene* began sailing between Trenton and Port Severn. Because of the time it took to construct, the Trent-Severn Waterway never achieved its aim of aiding the economic development of the Canadian part of the Great Lakes. However, the inaugural trip by pleasure boat was indicative of the future and today the waterway is an important part of Canadian tourism. In order to overcome the changes in level along the waterway, there are today 39 locks, two lock staircases, two hydraulically operated vertical

Fig. 81 The original Big Chute Marine Railway.

Fig. 82 Location of the Trent-Severn Waterway.

lifts and an inclined plane called the Big Chute Marine Railway, rise 58 feet (17.68 metres).[58]

At the time the waterway was opened there were two inclined planes, both with hydro-electric power plants, which made a passage of the Trent-Severn Waterway possible. As well as the Big Chute Marine Railway there was one, rise 47 feet (14.3 metres), for avoiding the river at Swift Rapids. Both were completed in 1916, though they were considered as temporary structures.[59]

Their design was typical of inclines that had been used up till then and such as have already been described. Boats weighing up to 20 tons were transported on a railway carriage open at either end and running on a single-track and which was controlled by ropes connected to an electrically-powered capstan.

In 1965, the Swift Rapids Marine Railway was replaced by a chamber lock. The Big Chute Marine Railway should also have been replaced

Fig. 83 Big Chute Marine Railway as it is today.

but for ecological reasons it has been retained. With an open waterway and lock, the parasitic sea lampry could have escaped upwards into the Lake Simcoe fishery. Instead, in 1977, a larger Marine Railway was built, which could take boats up to 100 feet (30.5 metres) long, 24 feet (7.3 metres) wide and 6 feet (1.83 metres) deep, similar to those which could use the locks, with a load capacity of up to 100 tons.[60] The old incline survives and is occasionally put into service.

The Czech Republic

The 1960s saw the construction of more dry inclines. One was built for pleasure boats weighing up to 3.5 tons on the River Moldau at the Orlik Dam. The dam, which opened in 1962, has a rise of 71.5 metres, the highest on the river. The carriage for carrying boats is raised to the summit where it rotates on a turntable before lowering back into the water. In this way, no alteration is needed to the slope of the track to keep the carriage level on the inclines on either side of the summit. Beside the pleasure boat incline, there should have been a second incline for boats of 300 tons carrying capacity.[61] This incline was never completed and in design belongs to the next group of lifts, namely to longitudinal inclines with wet operation.

Fig. 85 The carriage for pleasure boats sits on its turntable at the summit of the Orlik incline. Behind is the top of the uncompleted incline for 300 ton boats.

China's incline

Another incline with dry operation was opened in China in 1973 at the Danjiangkou Dam on the River Hanjiang. It works together with a vertical lift which is described on page 88. The two lifts together raise 150 ton boats 68.50 metres, the incline accounting for 33.50 metres. Boats up to 40 metres long by 10.70 metres width can be carried and the platform on which they sit can also be converted into a caisson, though this reduces the effective length to 24 metres and reduces the effective draught from 1.20 to 0.90 metres.[62]

Wet longitudinal inclines

To raise or lower boats in greater safety, the idea of transporting them in water-filled troughs or caissons had originally been made in 1830, Brennecke[63] suggests, by de Solages. Where the carriage of boats in a caisson on an incline first occurred cannot be ascertained exactly. The incline at Wellisford, rise 81 feet (24.7 metres), on the tub-boat section of the Grand Western Canal has already been mentioned and could have been the one. However, it is not possible to be certain beyond doubt.[64]

The Chard Canal

There are three other inclined planes which are early examples and all, as mentioned earlier,

Fig. 86 A recent view of the inclined plane at Ilminster.

are on the Chard Canal; Thornfalcon, rise 28 feet (8.53 metres), Wrantage, rise 27.5 feet (8.38 metres), and Ilminster, rise 82.5 feet (25.15 metres). On each of these inclines, completed in 1841/42, ran two six-wheeled, water-filled caissons, 28.6 feet (8.7 metres) long and 6.9 feet (2.1 metres) wide. The gates of the caissons opened inward and the system operated by adding water to the upper caisson. The slope of the first two inclines was 1:8 while Ilminster was 1:9. Boats on the canal carried eight tons, and were 26 feet (8 metres) long, $6^1/_2$ feet (2 metres) wide and $2^1/_4$ feet (0.7 metres) deep. Due to competition after construction of a railway parallel to its line, the canal was in receivership by 1853 and completely closed by 1867/68.[65]

Fig. 87 Drawing of the lower end of the Ilminster inclined plane from 1852. The water wheel provided power for the incline.

The Monkland Canal

Better known than the inclined planes on the Chard Canal was that on the Monkland Canal at Blackhill in Scotland. Boats on the canal were 66 feet (20.12 metres) long and 13.5 feet (4.11 metres) broad. Empty ones were raised 96 feet (29.26 metres) by the incline. Loaded boats carrying 70 tons of coal or iron ore down the canal continued to use the adjacent flight of locks.

Just $12^1/_4$ miles long, the Monkland Canal joins the Forth & Clyde Canal in Glasgow. It runs in a generally southwesterly direction to Monkland (so called because here in the Middle Ages monks found coal and iron ore), near Coatbridge. The canal was built at the end of the eighteenth century and at Blackhill there were four pairs of two-rise staircase locks. Increases in traffic forced the construction of a second set of locks at Blackhill around 1840 and resulted in an acute water shortage, despite reservoirs containing about 300 million ft^3 of water. Then, in 1849, the canal's engineer, James Leslie, revived an earlier suggestion that an incline should be built at Blackhill and this was completed in 1850.[66]

The incline was built with twin caissons, and both they and their carriages were made from iron. The caissons were sealed by steel lifting gates, supported by counterweights on chains and moved by cranks. The weight of each caisson was 70 tons which was carried on 20 wheels. The carriage was 7 feet (2.13 metres) higher at one end than at the other to ensure that the caisson was horizontal. The carriages were connected to the winding drums by two inch diameter wire ropes which were fastened to their lower end. Because of the counter-balancing effect, the caissons only needed power to set them in motion, to overcome friction and to cope with the weight variation of the ropes as they wound in or out. Power came from two high pressure steam-engines at the upper end of the incline and they could also be used for back-pumping into the upper canal.

This contemporary report of the operation of the whole plant has been taken from the magazine *The Engineer and Machinist* of July and August 1850:

The improvement now being introduced on the Monkland Canal, by Mr Leslie C E, and of which the following is a short description, is not intended to supersede the use of the locks, as has been asserted in various professional papers, but to save time and water, by taking up the empty boats, the loaded traffic on the canal being almost all downwards.

The main feature of the improvement is the inclined plane, up which the empty boats are pulled from below to the higher level. The lift is 96 feet, and the gradient being one in 10, and the length of the carriage 70 feet, the whole length of railway is consequently 1030 feet.

The boats, which measure 66 feet by 12 feet 8 inches, and draw from 18 to 20 inches when light, are taken up afloat in a wrought iron caisson, which is raised up to a considerable height above the rails at the lower end, so as to be level on a carriage having 10 pairs of wheels. The caisson is run down into the water of the lower reach of the canal, the boat floated in, and the gate or sluice shut, so as to confine the water. The caisson, with the boat and water, is then hauled up close to the gate of the upper reach of the canal, and a water-tight joint being formed between them, the caisson serves as a lock, so that the boat floats out of it into the upper reach, and the water flows in to replace it.

There are two lines of rails and two caissons, so that one is ascending with the empty boat while the other is descending filled with water. The rails have a 7 feet gauge, and the two lines are 18 feet 3 inches from centre to centre. The wheels, which are like those of an ordinary railway carriage, are 3 feet diameter, except to two upper pairs, which are respectively 2 feet 3 inches and 18 inches diameter.

There are two coupled high-pressure engines at the upper end of the incline, driving two drums of 16 feet diameter on separate shafts, and work-ing in opposite ways, so that in both the rope is over the top of the drum. The rope is of wire, two inches diameter. The weight of the carriage, caisson, water and boat, is about 70 tons, and the time of ascent will be about five minutes.

The scheme of the inclined plane was first proposed by the late Mr Andrew Thomas, C E, in 1839; but his plan had only one line of rails, the other being occupied by water vessels, which were to be the moving power. In that year, on being consulted as to Mr Thompson's plan, Mr Leslie proposed a modification of it, bringing it to be nearly similar to the one now adopted, except that the caisson was to be of timber, and a chain was to be used instead of a wire rope. He also suggested another plan for bringing up the vessels on an open cradle, so as to save the weight of water; but that was objected to, from fear of injury to the boats.

In 1840, Sir John McNeill reported on, and recommended the adoption of, a somewhat similar plan, worked either by steam power or by a water wheel; but, instead of adopting any of the plans proposed, the Monkland Canal company built a second set of locks, which sufficed for all their trade, until last summer, when the water ran short, and the canal was shut for six weeks.

Mr Leslie was then consulted by the Forth & Clyde Canal company, to whom the Monkland canal now belongs, along with Mr Bateman, their engineer, as to what ought to be done to save water, or to provide the necessary supply; and they agreed in recommending the adoption of the plan formally proposed of the inclined plane, with modifications.

The caissons and machinery were contracted for by Messrs Yule and Wilkie, of Glasgow, and the work is now nearly completed.

Originally it was only intended to use the incline in times of water shortage in summer. However, the incline operated so well that this was forgotten and it worked throughout the year. In place of the 30 to 40 minutes taken in passing through the locks, the time for the inclined plane was only about eight minutes. Between the 26th March and the 3rd September 1853, some 6,456 boats ascended the incline and 184 descended, with around 7,500 boats being handled annually between 1852 and 1856, with a peak of 8,674 boats. The water consumption per operation was about 50 ft³.[67] The caisson at Blackhill was later converted to dry operation because surging of water in it could not be controlled sufficiently. This is a major problem on inclines with longitudinal caissons.

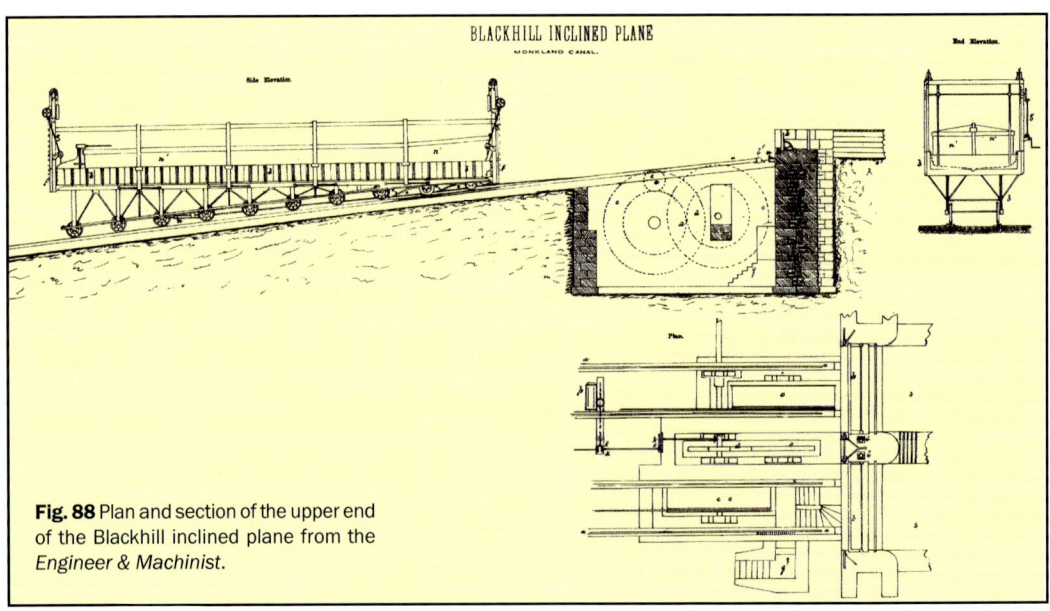

BLACKHILL INCLINED PLANE
MONKLAND CANAL.

Fig. 88 Plan and section of the upper end of the Blackhill inclined plane from the *Engineer & Machinist*.

66

Fig. 89 Location of the Monkland and Union canals.

Map labels: Falkirk Wheel, Forth & Clyde Canal, Falkirk, FIRTH OF FORTH, Edinburgh & Glasgow Union Canal, Edinburgh, River Clyde, Monkland Canal, Coatbridge, Glasgow, Blackhill Locks and Inclined Plane

1 Kelvin Aqueduct
2 Falkirk Tunnel
3 Avon Aqueduct
4 Almond Aqueduct
5 Slateford Aqueduct

The maximum volume of traffic on the incline was achieved in the 1860s. For example, in 1863, 1.5 million tons were handled by the incline.[68] The end of the inclined plane at Blackhill came sooner than expected because of competition from railways.

Over the years after 1866, traffic declined to half previous figures and because the cost of operating the steam-engines, with the reduced use of the incline, was uneconomic, it was taken out of service around 1887 and once more all traffic passed through the locks.

The last coal was carried in the 1930s and most of the bed of the canal was removed during construction of a motorway. It is now difficult to find the location of the incline. However, in its time it created great interest and for his design Leslie received the Silver Medal and Plate of the Royal Scottish Society. Towards the end of its working life the incline was visited by French engineers who were building the one on the Canal de l'Ourcq and it is even suggested that it inspired the Belgian engineers who were designing the incline at Ronquières, which will be described later.[69]

The Georgetown incline

The next incline with a longitudinal caisson was erected in the USA, at Georgetown on the Chesapeake & Ohio Canal. Construction of the canal began in 1828 though it never reached the Ohio. It was nevertheless very important for the transport of coal from the Cumberland mines.

When a record annual tonnage of almost one million tons of coal were carried, boats had been held up whilst waiting to descend the locks into the Potomac River. To relieve the problem, in 1875 it was decided to augment the existing locks with an inclined plane.

On the 600 feet-long incline, which came into service on 29th June 1876, there was a single wooden caisson, 112 feet (34.13 metres) long, 16.75 feet (5.11 metres) broad and 7.8 feet (2.38 metres deep), which was carried on six six-wheeled bogies running on two parallel tracks. The weight of the caisson was balanced by two counterweights which ran on separate tracks, each 300 feet (91.4 metres long) to the right and left of the main track. Power to overcome friction and other losses was supplied by a water turbine. The rise between canal and river was 40 feet (12.2 metres) at normal river levels and the slope of the incline was 1:12.[70]

The operation of the incline, on which boats of 125 tons, 90 feet (27.4 metres) long, $14\frac{1}{2}$ feet (4.39 metres) wide and 5 feet (1.52 metres) draught were carried, was something of a failure. The construction, in particular the drive ropes, track and masonry, were not initially suitable for the substantial weights encountered. The weight of the water-filled caisson was about 400 tons and the counterweights were 200 tons each. After several small incidents, on the 30th May 1877 there was a spectacular accident. The masonry which supported the pulleys gave way, breaking the ropes and causing the caisson and its counterweights to roll out of control into the river. The deputy harbour master, John Mead, was killed and two of his helpers severely

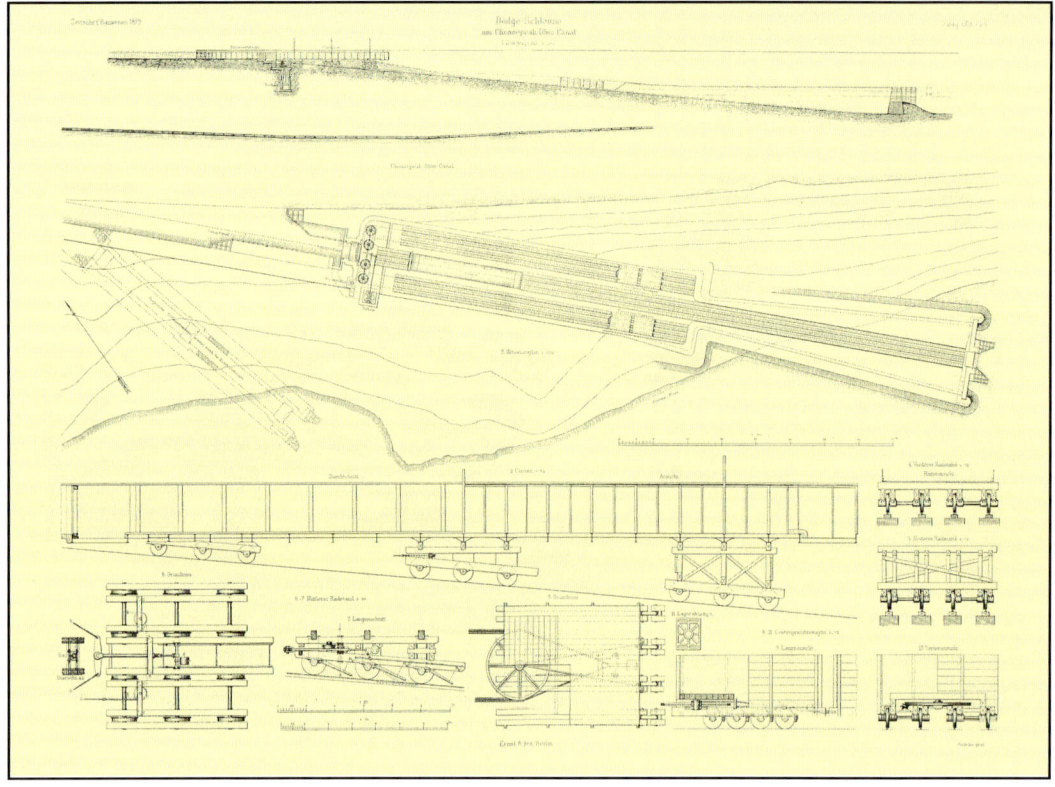

Fig. 90 The inclined plane at Georgetown on the Chesapeake & Ohio Canal.

injured. The caisson and the boat within were not seriously damaged and the boat's crew were unharmed.

The chief engineer responsible for building the incline, William Rich Hutton, improved the brake system because of the accident and replaced the pulleys by heavier ones as well as changing the iron rails to steel. Subsequently, water was drained from the caisson after a boat had entered to reduce the load on the system.[71]

Worse than these teething troubles was that traffic on the canal was declining. In 1877 only 627,913 tons were carried[72] and in 1878 just 1,918 boats used the incline, about seven per day.

The end for the incline came abruptly with the great Potomac flood of the 30th May to the 1st June 1889, when part of the canal as well as the incline were destroyed. A little masonry survives from the incline, which was used by the USA at the Paris World Fair in 1878 as an example of American structural and water engineering.

The proposed Danube-Oder Canal

After the incline at Blackhill was taken out of service and the destruction of the incline at Georgetown, there were no further longitudinal inclines with caissons until 1968 when two huge inclines came into service, in Belgium and in the former Soviet Union. However, the idea of such inclines did not disappear completely in the interim. In 1903, the Austrian government held a competition for the design of a lift on the 35.9 metre rise on the planned Danube-Oder Canal at Prerau (Prerov). Designs of many different kinds were considered by the jury and they awarded the first prize to a Czech/Austrian consortium.

This 'Universell' design proposed using a longitudinal incline with twin operation. It was to have a slope of 1:25 with two tracks, gauge 6.30 metres, and a geared drive rack lying between them. The caissons were to have a water depth of 2.30 metres and were for boats 68 metres long and 8 metres broad. They were to be powered by two electric motors fitted beneath the caisson with a gear drive engaged with the rack in a similar fashion to mountain railway locomotives. The carriages were not to be connected mechanically, by ropes or rods, but would be kept in balance by an electrical linkage. Each carriage would have a dynamo/motor in phase with the main drive motor. The counter-weight effect was to be generated by connecting the ascending dynamo/motor with the descending one. The electricity generated during descent was to be supplied to the other as it ascended and this would have helped or hindered the operation of the main motor, depending on the relative speeds of the two carriages. Everything was to be controlled from the carriage and they could each be worked independently if necessary.[73]

Unfortunately the Danube-Oder Canal has not been built despite the efforts of Czech authorities and economists, particularly after 1945, and so the incline at Prerau and others on the proposed canal remain just a dream.

Further German proposals

The same can be said for the Neckar-Danube Canal and the Danube-Lake Constance Canal, inclines which were designed by Konz together with the Dortmunder Union Brückenbau A.G. and the Siemens-Schuckert-Werke A.G. On the first canal there would have been two double inclines of about 100 metre rise and a slope of 1:21, and on the second waterway an incline of 104 metres rise and a slope of 1:47.5. This would have resulted in a track almost 5,000 metres long.[74] The canals were to be for 1,200 ton vessels, 80 metres by 10.25 metres by 2.30 metres. Their construction was not undertaken because of changing transport requirements.

Krasnoyarsk incline

This incline was erected between 1962 and 1968 on the left bank of the Yenisey to bypass a new hydro-electric dam. It was decided not to

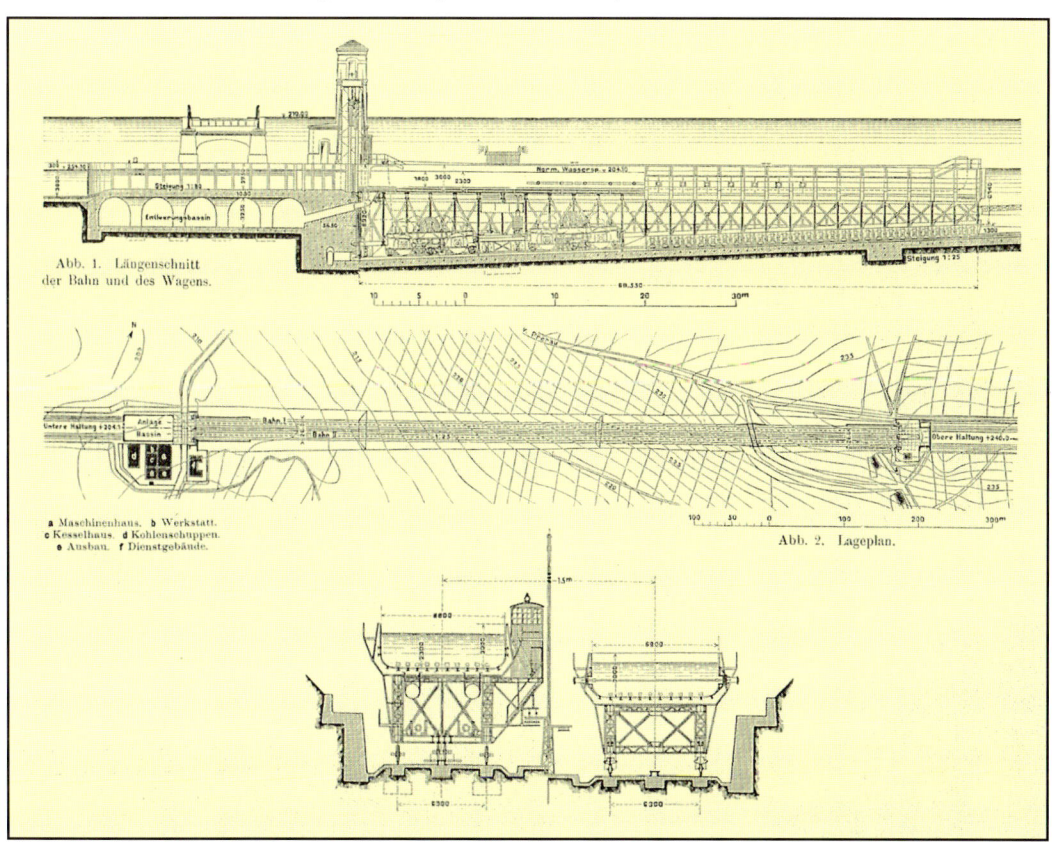

Fig. 91 The competition-winning 'Universell' inclined plane design for the Danube-Oder Canal.

69

Fig. 92 & 93 Two views of the Krasnoyarsk inclined plane.

use a lock staircase because of the rise of over 100 metres, the large variation in water level, upstream of up to 13 metres and downstream of up to 6.3 metres, and because of the difficult terrain.[75]

The incline consists of four main parts: two inclined planes with trackways into the upper and lower waters, the turntable linking the two inclines and the self-propelled caisson.

The caisson has a usable length of 90 metres and width of 18 metres and with a water depth of 3.30 metres, its total weight is 6,720 tons and can take boats of up to 2,000 tons. Its carriage has two sets of 39 pairs of wheels, which run on a track of 9 metres gauge.

The caisson is provided with a movable segment gate at its head and this is lowered for opening. At the closed end of the caisson are the operators' cabin and electrical machinery and the main drive machinery is housed along the side of the caisson. A hydraulic drive was chosen as the best solution, consisting of 18 axial reciprocating pumps and 156 radial piston hydraulic motors, each of 75 kilowatts. The shaft of each hydraulic motor has a brake and these can hold the caisson on the incline after it has stopped and can also be used if there is a breakdown. On the ends of the hydraulic motor shafts, whose axes are vertical to the slope of the plane, are fitted drive cogs of 1.05 metres

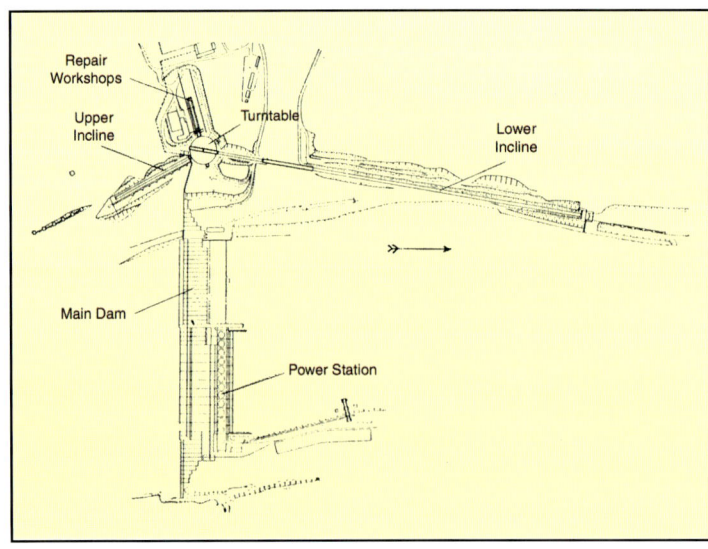

Repair Workshops

Upper Incline

Turntable

Lower Incline

Main Dam

Power Station

Fig. 94 Layout of the Krasnoyarsk inclined plane.

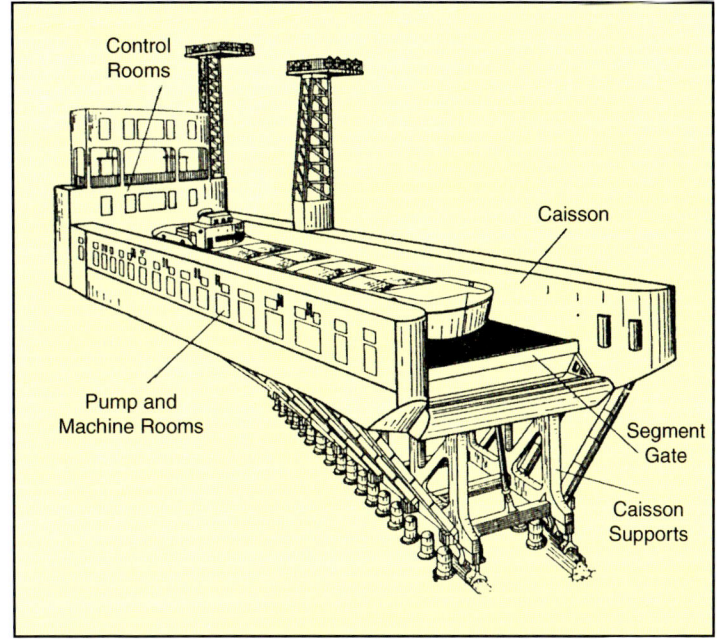

Control Rooms

Caisson

Pump and Machine Rooms

Segment Gate

Caisson Supports

diameter. These engage in double-tooth racks and provide the drive while ascending and braking during descent.

The length of the inclines is 306 metres from the summit to the upper level and 1,189 metres to the lower. They lie at an angle of 142 degrees to each other and meet either side of the 104.8 metre turntable. The slope of the planes is 1:10.

The uphill speed is 1.00 m/s and downhill 1.33 m/s and a complete cycle, both ascent and descent, requires 93 minutes, so that up to 15 operations a day are possible.[76] With a rise of 100 to 108 metres, the Krasnoyarsk incline is the world's largest lift.

Ronquières incline

This incline on the Canal de Charleroi à Bruxelles took six years to build and opened in 1968. The canal has linked the industrial area around Charleroi with the Belgian capital since 1832 and from there via the Brussels-Rupel Canal, the Rupel (a tributary of the Schelde) and the Schelde with the port of Antwerp.

The canal from Seneffe, on the Canal du Centre, to Brussels was built for 70 ton boats and had 38 locks and a 1,267 metre long tunnel. It had been improved to 300 ton standard by 1914 and after the First World War enlargement for 1,350 ton boats was begun. This was completed when the Ronquières incline was opened. Eight large locks and the incline replaced the 38 small locks. The time taken for travelling between Charleroi and Brussels was shortened from 25 hours (for a 300 ton boat) to 18 hours (for a 1,350 ton boat), greatly increasing the amount of traffic the waterway could carry.[77]

Fig. 96 The Krasnoyarsk inclined plane, with the turntable in the foreground.

Fig. 97 The upper entrance to the Ronquières incline.

Fig. 98 A view down the Ronquières incline.

The incline is designed as a double lift with a 1,430 metre long track and overcomes a rise of about 68 metres. It was designed by Professor Willems, of the Free University of Brussels, after economic comparisons between the cost of four locks of 17 metre lift, two 34 metre vertical lifts and a single 68 metre vertical lift.

The caissons run on 1:20 inclined tracks and have a usable length of 87 metres, a width of 12 metres and a water depth which fluctuates according to the water level in the upper canal of between 3.00 and 3.70 metres. A caisson has a total weight of between 5,000 and 5,700 tons and is carried on two rows of 59 fully sprung axles with steel wheels running on a railed

track. Each caisson has 8 traction cables of 55 mm diameter run over guide pulleys and a drive drum of 5.50 metres diameter. This is powered by 6 DC motors, each of 125 kilowatts, which are located at the upper end of the plane. The cables are connected to counter-balances which comprise wheeled counterweights of 5,200 tons running under each caisson.

The caisson gates as well as the gates to the upper and lower canals are raised or lowered from gantries over the end of both planes. The normal speed of a caisson is 1.20 m/s, and the time taken to ascend or descend the incline is around 22 minutes. To avoid surging in the caisson, acceleration and deceleration are both

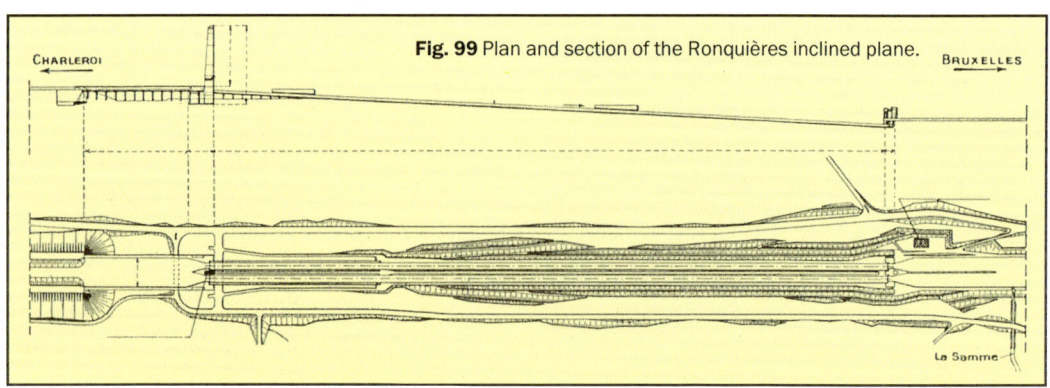

Fig. 99 Plan and section of the Ronquières inclined plane.

Fig. 100 One of the caissons on the Ronquières incline.

carefully controlled. Boats are also provided with flexible moorings for this reason.

The incline and its surrounds have been developed for tourism and the best view of the incline, from where its full length can be seen, is from the top of the 125 metre high tower at the upper end of the incline. Unfortunately it is closed to the public at the moment.

Transverse inclined planes

When the location for an inclined plane has a greater slope than 1:15, a longitudinal caisson and its carriage become too expensive and other alternatives must be sought. The use of a transverse caisson is possible on steeper slopes but they are expensive and are only used where

the slope is between 1:2 and 1:8. For slopes between 1:8 and 1:15, neither transverse nor longitudinal caissons provide economically satisfactory results and in this case, as Simons suggests, you should use 'the lesser evil'. When the slope is steeper than 1:2, then only a vertical lift will provide the answer.[78]

On transverse inclined planes, boats are not moved in their normal direction. As canals are usually laid out parallel to contour lines, it is better to conform to this direction of travel. Transverse inclined planes have up to now only been designed and built for use with water-filled caissons.

The Foxton incline

For a long time, the only lift with transverse caissons was that at Foxton in Leicestershire, England, which was built at the beginning of the twentieth century.

The Foxton incline is on the old Grand Union Canal. Along with the Leicester Line, it was taken over in 1894 by the Grand Junction Canal Company whose canal had been built between 1793 and 1805. It became part of a new Grand Union Canal Company in 1928 to create the longest canal system in Great Britain. This has

Fig. 101 Location of the Foxton inclined plane.

Fig. 102 The Foxton inclined plane shortly after it opened to traffic.

a 137 milc (220 km) main line from Brentford Lock on the Thames to Birmingham with 166 locks and a 77 $^5/_8$ mile (125 km) branch with 73 locks from Norton Junction to Langley Mill via Leicester. From Brentford to Birmingham, the canal was broadened in the 1930s, while that to Leicester was built for wide boats, though with narrow locks at Watford and Foxton to save money. Two sets of five-rise locks were built at Foxton with a total rise of 75 feet (22.9 metres), only separated to create a passing place.

In July 1896, because of the poor condition of the locks at Foxton, the Grand Junction Canal Company decided to build an incline with a transverse water-filled caisson, the world's first. A large scale working model was built at the canal company's works at Bulbourne to test the system and, after improvements to the design, construction began at Foxton in 1897. The incline came into service on the 10th July 1900. Each of the two caissons has a length of 80 feet (24.40 metres), a width of 15 feet (4.57 metres)

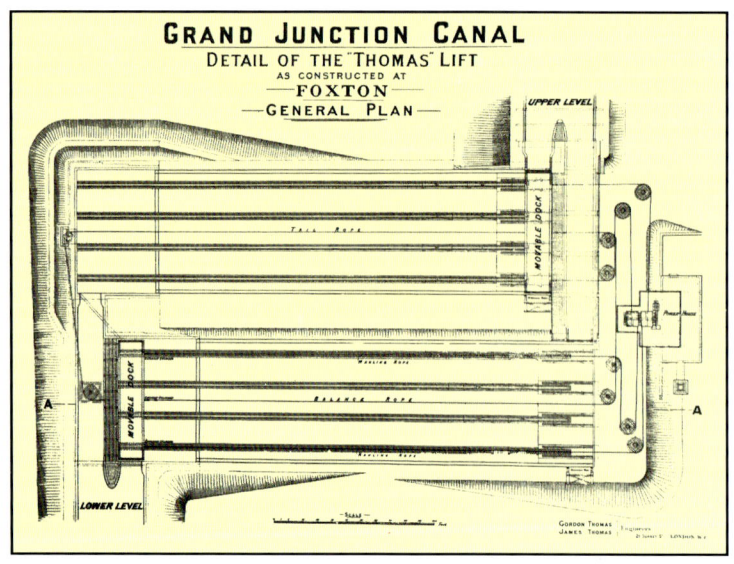

Fig. 103 Plan of the Foxton incline.

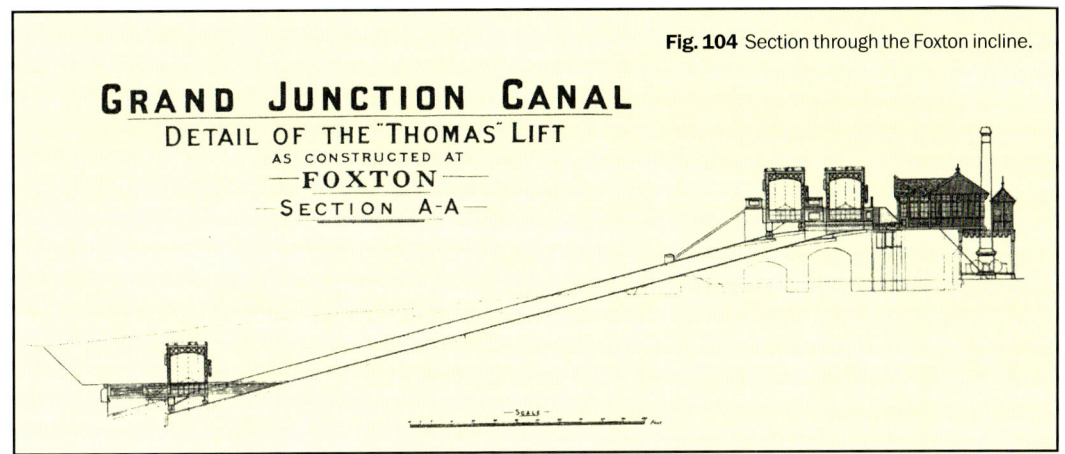

Fig. 104 Section through the Foxton incline.

GRAND JUNCTION CANAL
DETAIL OF THE "THOMAS" LIFT
AS CONSTRUCTED AT
—FOXTON—
—SECTION A-A—

—SCALE—

and a water depth of 5 feet (1.52 metres). The 1:4 incline was designed as a twin lift, taking either a 70 ton wide boat or two narrow boats carrying up to 33 tons each. The caissons, fitted with eight pairs of wheels running on four double lines of rails, were connected by four wire ropes. At the lower level they entered the canal so that here there were no gates. The caissons were fitted with lift gates at both ends which were balanced by counterweights and lifted by hydraulically operated cylinders. At the upper level, similar lift gates were fitted and the caissons were held against the gate frames, which were built into the end wall, by hydraulic pressure. In this way a water-tight seal between the caisson and upper canal was made.

The time for traversing the 315 feet (96 metre) long incline was about 4 minutes and two boats could pass through the system in 12 minutes in contrast to the 80 minutes taken through the old lock staircase, a great improvement.

Fig. 105 The author inspecting the inclined plane.

Fig. 106 The winding drum and the 25 horse-power steam engine, used to overcome friction and to provide hydraulic power.

75

The incline was closed after just ten years operation, not for technical reasons, but more because of financial problems. It was not worked at night, because it was uneconomical to keep the boilers in steam. This meant that the locks had to be improved and returned to service. Three extra lockkeepers were needed, adding to the expence of the operation. Because of this, the canal company decided in October 1910 to discontinue the routine operation of the incline. In March 1911 it was closed, the beginning of its slow death. It survived intact for many years but was used only during lock repairs. After the First World War it was decided to scrap it but several more years were to pass as the company waited for better scrap prices. The incline was finally demolished in 1927.[79]

In 1980 the Foxton Inclined Plane Trust was founded with the object of both recording and conserving the memories of this once unique incline. In the surviving engine house, a small museum has been established with displays about the incline and the English waterway system. The site has been cleared by members of the Trust and the incline is still recognisable and some parts of the upper walls have survived. Altogether the area around the locks and old incline is very interesting and well cared for so it is not surprising that it has become one of Leicestershire's main tourist attractions.

The Arzviller incline

In 1969, at Arzviller on the Rhine Marne Canal, the second and, up to now, the only other transverse inclined plane opened. The canal, which joins Vitry-le-François, on the Marne Lateral Canal, to Strasbourg was opened in 1853 some 15 years after construction began. The 314 km long canal originally had 178 locks and two summit levels, between the valleys of the Meuse and Meurthe, and the Meurthe and Zorn.

The Arzviller incline is located where the canal descends from the northern edge of the Vosges to the upper Rhine plain and where, in a distance of just four kilometres, there were seventeen locks to overcome a change in level of 44.45 metres. The average distance between locks here was only 180 metres and in two cases just 45 metres. Boats carrying up to 280 tons took about 8 to 10 hours to pass through the locks.

It was during the enlargement of the canal to 350 ton standard that this flight of locks was replaced by a transverse inclined plane. There was an international competition for the design, nine firms submitting altogether 40 suggestions

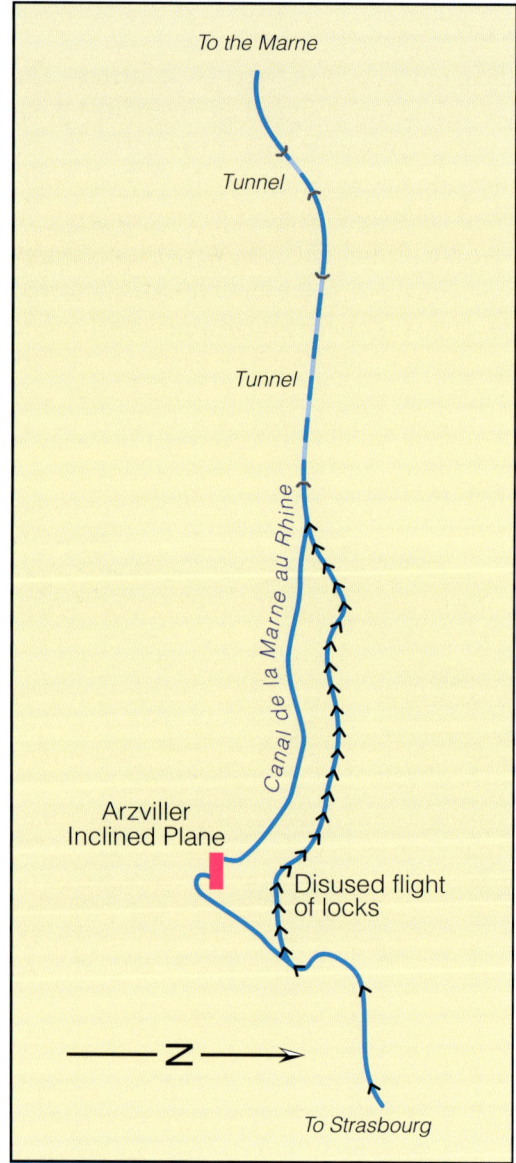

Fig. 107 Location of the Arzviller inclined plane.

Fig. 108 The drive machinery and brake gear on the Arzviller inclined plane.

metre diameter sprung rollers at the centre of the caisson control sideways movement and run on rails fixed on top of a triangular concrete support. There are two counter-balance weights which run on rails below the caissons and are connected to them by straps consisting of 14 cables carried over a guide pulley. Because of varying water levels in the canal, it is possible for the weight of caisson and counterweight to become slightly out of balance. The drive and control mechanisms for the incline are all installed at its upper end.

The incline was built with a single caisson, but was designed such that, by extending the upper and lower canal entrances, a second caisson with a higher undercarriage could be installed, allowing the incline to function as a double lift.

Both the caisson and upper and lower canals are fitted with lift gates. When the caisson is at its upper or lower position, the space between the caisson and the canal entrance is sealed by a movable frame. When this is in place, the caisson and the canal gates can be lifted so that boats can enter or leave. Only the gates on the entrance canals are fitted with a lift mechanism. These also raise the caisson gates which can be attached to them by means of hooks.

After a boat has entered, both the caisson and the canal gates are lowered simultaneously and

including longitudinal and transverse inclines, water slopes and vertical lifts, from which the best was chosen.

A transverse incline was considered most suitable for the steep terrain and it also allowed for construction and testing without interfering with the operation of the old canal locks. It has a rise of 44.55 metres, a length of 136 metres and an inclination of 1:2.44. The caisson is 41.50 metres long, 5.50 metres wide and has a water depth of 3.20 metres. It weighs 894 tons. The caisson undercarriage has two four-axled bogies at each end which run on two pairs of rails set 25.75 metres apart. Each pair of bogies is linked to two 3.30 metre diameter winding drums which are driven by two 100 kilowatt electric motors. Since the system is in balance, small electric drive motors are sufficient. Two

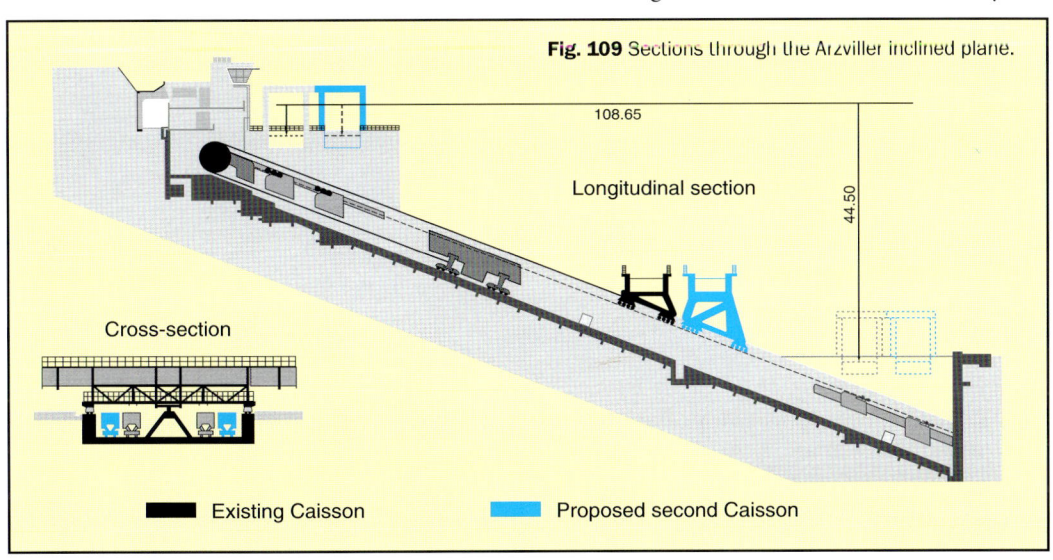

Fig. 109 Sections through the Arzviller inclined plane.

108.65

Longitudinal section

44.50

Cross-section

■ Existing Caisson   ■ Proposed second Caisson

Fig. 110 The Arzviller inclined plane in operation.

the space between the gates is emptied by opening a slide valve. After emptying, the safety hooks and seal are released and the caisson is free to move.

The acceleration and deceleration of the caisson is 0.02 m/s² and its maximum speed of 0.6 m/s is reached after 12 metres. The time taken from top to bottom is about 4 minutes and a complete passage up or down takes less than 20 minutes. If the lift is open for 13 hours, 39 cargo boats (19 in one direction and 20 in the other) can pass each day. The incline opened on the 27th January 1969.[80]

The number of commercial boats using the canal has reduced considerably in recent years, while pleasure and recreational traffic has increased. The incline has become, as its many visitors show, a major tourist attraction. From the upper end of the incline this impressive lift can be seen at its best. During the summer months there is a small passenger boat which offers regular trips that include a passage of the lift. Nearby a small visitor centre has been set up in a péniche, which gives an insight into

boating life and the history of the construction and operation of the lift.

Since Arzviller opened, no more transverse inclines have been built, but there have been some interesting developments in France.

Water-slope inclines

Contrasting with the more obvious ways of building a lift, in which today the boat is usually transported in a caisson, the water-slope incline uses a movable gate which fits into a sloping concrete channel constructed between the two water levels. In this, a wedge of water in which the boat floats is constrained to move upwards or downwards by the gate. It operates as follows. After a boat enters from the lower canal into the wedge of water at the foot of the concrete channel, the movable gate is lowered behind it. By pushing the gate upwards along the channel, the wedge of water and the boat are raised to the upper level. The upper canal level is closed off by a fixed gate until the pressure between the water in the wedge and that in the upper canal is equalised. The upper gate can then open and the boat sail out.

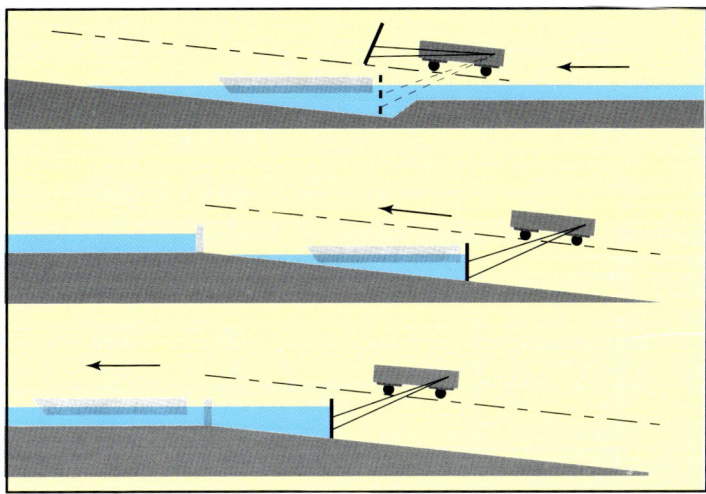

Fig. 111 Method of operation of a water-slope inclined plane.

Water-slope lifts are only suitable for slopes of around 1:30, as with steeper slopes the useful length of the water wedge becomes too short to accommodate a boat. The most favourable rise for such inclines is from 14 metres to a maximum of 40 metres. Larger rises are not possible because of the length of channel which would be required.

The main advantages of a water-slope lift, as described by Partenscky, are:

1   an extremely favourable relationship of payload to total load, as a caisson is not required,
2   considerable cost savings compared with locks or other lifts of equal rise,
3   practically no increase in journey times, as a boat in the water wedge is moved with a similar speed to that when in the canal,
4   push-tows do not need to be uncoupled but can be raised or lowered as a unit.

A disadvantage with such inclines is the difficulty of sealing the gate against the channel and the severe wear to which the seals are subjected. Holding boats in position against the gate and tractor can also cause problems and, despite what Partensky suggests regarding push-tows, there is a physical restriction on the length of boat or unit which can use a water-slope. For a modern 196 metre push-tow, one end of the wedge of water on a 4% gradient would need to be about 12 metres in depth. In practice, water-slopes are only really suitable for single barges.[81]

A German design

A scheme for water-slope lifts was first suggested in the late nineteenth century. In a technical book by the German engineer Julius Greve, published in 1885, he sets out the basis for such an incline. It is interesting that Greve's design was for use by standard large Plauer boats (67.00 m long, 8.20 m broad, 2.00 m depth, 760 ton load capacity) and had a rise of 25 metres. It was designed with two parallel channels, the gate in each being connected by a chain passing around a wheel at the upper end out of the water, thus creating a twin lift. To operate the system, the downward-moving wedge was to be overfilled by 0.35 metres. This would create a force of 18 tons and calculations suggested it would result in a speed of 1.5 m/s. The overfilling would have required 275 m^3 of water for each operation.[82] Greve's idea was never put into practice and it was only in mid-twentieth century France that the water-slope lift became a reality.

Montech and Fonserannes water-slopes

It is thanks to Aubert's fundamental work that water-slopes were built at Montech on the Canal latéral à la Garonne in 1973 and at Fonserannes on the Canal du Midi in 1983.

Opened in 1856, the 193.6 km long Canal latéral à la Garonne has 53 locks and is part of

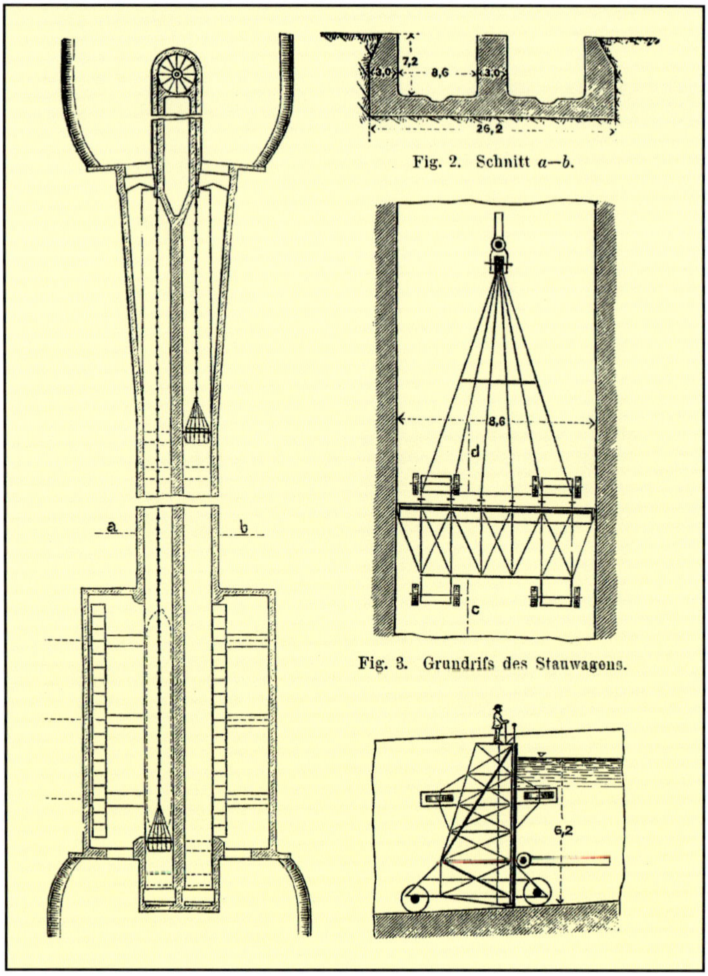

Fig. 112 The water-slope which was designed by Julius Greve.

Fig. 2. Schnitt *a—b*.

Fig. 3. Grundrifs des Stauwagens.

the waterway between the harbour of Sète on the Mediterranean and the Atlantic to the north of Bordeaux. The locks are shorter than the Freycinet standard and to modernise the canal between Toulouse and Castets en Dorthe, 50 km south-east of Bordeaux, it was decided in 1968 to lengthen them. At Montech, where five locks close to each other overcame a rise of 13.3 metres, a water-slope, the world's first, seemed an attractive alternative. It would avoid the locks and be capable of taking a péniche, 38.5 metres long, 5.5 metres broad and load capacity 350 tons. The detailed preliminary research, including two models (the second to a scale of 1:10 and representing a 12 metre broad boat cross-section), confirmed the efficiency of the principle and resulted in the construction of the prototype water-slope.

The concrete channel has a slope of 1:33.3, a rise of 14.3 metres, a width of six metres and a wall height of 4.35 metres. It has to be accurately constructed so as to avoid the risk of the gate jamming. Two tractors, one on either side of the channel, move the gate. Each is powered by a 735 kilowatt (1000 HP) diesel-electric motor driving 4 axles (with eight rubber-tyred wheels), and both tractors are rigidly connected by a cross bracket, while the segmental gate, which can be moved up and down hydraulically, is fixed by an arm to this. In front is a horizontal U-shaped beam against which the boat is held. Side arms connect this shield protection beam with the tractors.

The upper canal is shut off by a hinged downward-folding gate. To transform safely the kinetic energy of the wedge of water as it

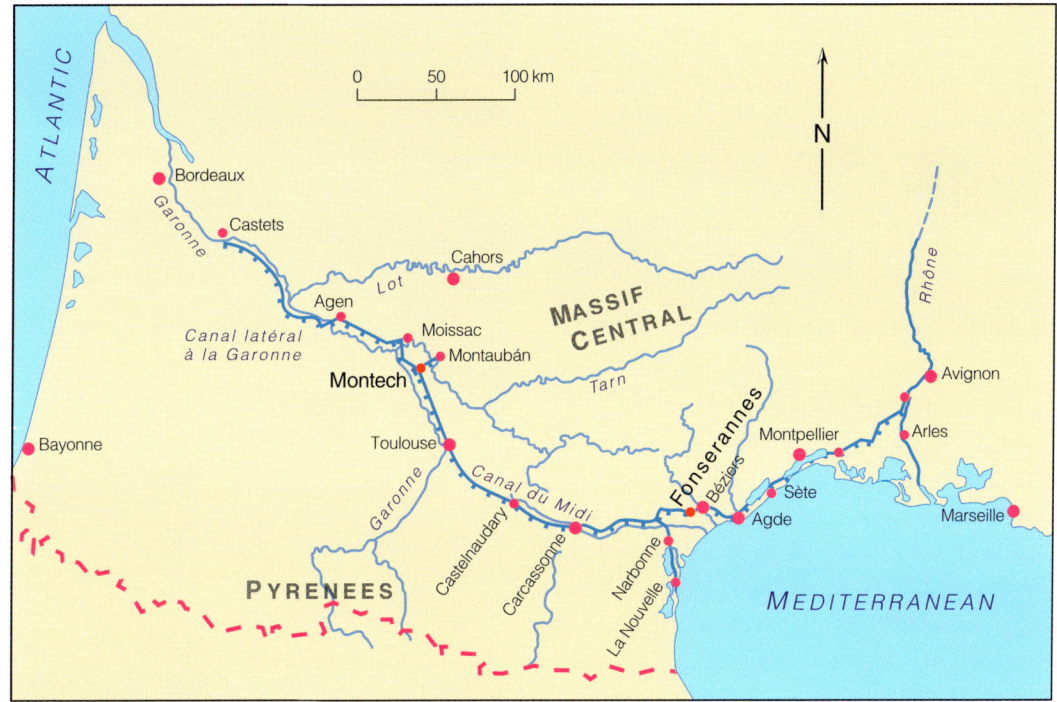

Fig. 113 Location of the Canal Latéral à la Garonne and the Canal du Midi between the Atlantic and the Mediterranean.

arrives at the top, there is a 27 metre long suppression canal, with herring-bone pattern ribs on its floor, between the upper end of the channel and the folding gate, with a settlement basin alongside. These measures prevent the excessive reflection of the waves produced by the water wedge on the folding gate.

The water wedge has a length of 125 metres and moves with a speed of about 5 km/h. The water depth in front of the gate is 3.75 metres, and the measured loss of water by leakage is negligible, from one to three litres per second depending on the condition of the neoprene sealing rollers.[83]

Unfortunately, the water-slope is fenced off, making it difficult to appreciate its method of operation from outside. Instead, travel up it on the passenger boat which operates on most days. Because there is no possibility for boats to moor, otherwise than bow or stern on to the shield protection beam, pleasure craft have continued to use the five locks.

Fig. 114 A péniche leaves the upper end of the Montech water-slope

Fig. 115 The energy dissipation channel at Montech.

The water-slope at Fonserannes on the Canal du Midi, which was constructed 10 years after Montech, is somewhat more 'tourist-friendly'.

The Canal du Midi was built between 1667 and 1681 to create a waterway link joining the Mediterranean (Gulf of Lyon) to the Atlantic Ocean (Bay of Biscay). Its exemplary design and construction inspired generations of canal engineers and visitors and still inspires today. Because of its importance to the history of waterway development, it has recently become a World Heritage Site.

The canal, from Port de l'Embouchure in Toulouse to Port des Onglous on the Étang de Thau, is 240.13 km long. The whole passage from Atlantic to Mediterranean, including the Garonne and the Étang de Thau, is about 600 km, as opposed to the sea route through the Straits of Gibraltar which is about 3,000 km. The Garonne is now partially by-passed by the Canal latéral à la Garonne.

The rise from the Garonne at Toulouse to the Col de Naurouze is 56 metres, a length of 51.6

Fig. 116 The water-slope at Fonserannes in operation, showing pleasure boats tied up to the movable mooring.

km with 17 sets of locks, of which nine are two-riser staircases. The 189 metre descent to the Mediterranean is 188.5 km in length and is overcome by 45 sets of locks, of which ten are two-risers, five three-rise, one four-rise and the then waterway wonder of the world, the eight-rise locks in Fonserannes.

To avoid lengthening the staircase locks at Fonserannes to Freycinet dimensions (38.5m by 5.5m), a water-slope was built in 1983. The slope is of similar width to that of Montech, with

Fig. 117 The tractors and gate of the Montech water-slope

Fig. 118 The tractor halfway up the Fonserannes water-slope.

Fig. 119 A view from the bottom of the water-slope at Montech.

Fig. 120 The movable gate raised. Note the sealing rollers on the side and bottom which run along the concrete channel when the gate is lowered.

a rise of 13.6m and a slightly steeper gradient, five per cent instead of three.

The water-slope failed shortly after opening, when a combination of hydraulic seal leakage and imperfect synchronisation of the massive wheels resulted in a serious incident. The traction unit skidded down its track, to the consternation of the occupants of two hire boats. It took until 1986 to resolve the technical, contractual and insurance issues. The water slope was then used for several years, but never to the satisfaction of the operator, who baulked in particular at the electricity bill. Re-opening is being considered because of the increasing numbers of pleasure boats, but substantial modifications will be required to make the structure both more reliable and less energy-intensive. Unlike the Montech slope, it has a

longitudinal beam with a set of bollards, which moves with the gate, so that small craft remain safely moored as they move up or down.

It is interesting to compare the projected costs of a high-capacity water-slope with a chamber lock. A lock with a comparable 20 metre lift is about 35% more expensive and a 29 metre lift twice as expensive as a water-slope.

VERTICAL LIFTS

With the boat out of water
Halsbrücke
Vertical lifts where boats are raised dry, out of water, are comparatively rare. However, their technology is simple and they are suitable for small boats, so it is not surprising that the earliest surviving vertical lift was designed in this way. It is near Halsbrücke in Saxony.

This lift, also known as the Rothenfurther Hebehaus, is on an abandoned small industrial

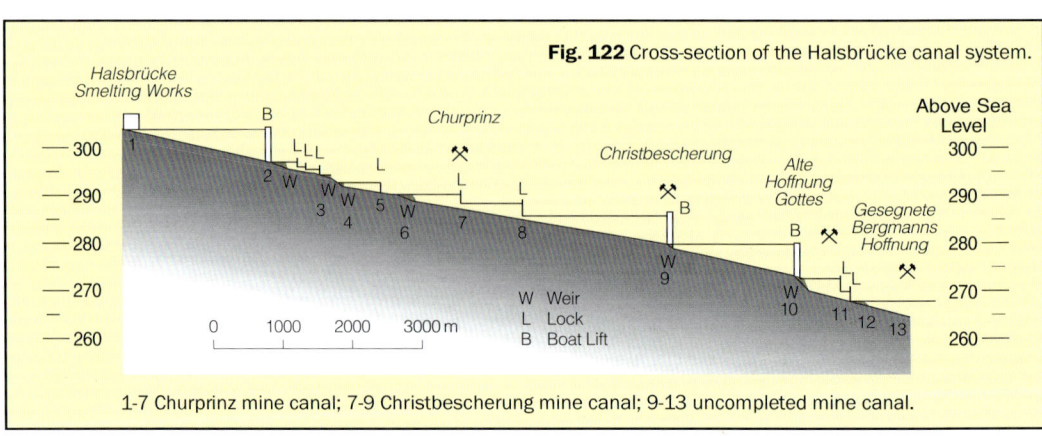

Fig. 121 A recent view of the boat lift at Halsbrücke.

canal. It had no connection with the national waterway network and, like similar English canals, worked in isolation. The reason for its construction was the cost of road transport, this being around eight times higher. The 5,350 metre long waterway was built between August 1788 and the autumn of 1789. Construction was under the direction of *Kunstmeister Mechanikus* Mende, who was also responsible for making the rivers Unstrut and Saale navigable from Artern to Weißenfels.

After the Churprinz Canal had been built, boats of 8.50 metres length and 1.60 metres width transported around $2\frac{1}{2}$ to 3 tons of ore from the Kurprinz silver mine at Großschirma to the smelting works further up the River Mulde and to Europe's then largest mercury works at Halsbrücke.

Each boat had a three man crew, of which two hauled from the land and one on board steered it by shafting … At least two boats would depart simultaneously from the terminus so that the crews could help each other at the lift and at the locks.[84]

The following is a contemporary description of the Halsbrücke lift by the German engineer Gotthilf Hagen, perhaps the most competent technical witness of the time:

I shall describe them as I have seen them in the year 1823 … The canal of about 13 feet (4.08 metres) width crosses the valley of the Mulde and climbs suddenly 18 Saxon yards

Fig. 122 Cross-section of the Halsbrücke canal system.

Halsbrücke Smelting Works

Above Sea Level

W Weir
L Lock
B Boat Lift

1-7 Churprinz mine canal; 7-9 Christbescherung mine canal; 9-13 uncompleted mine canal.

Fig. 123 The Christbescherunger Hebehaus or Grossvoigtsdorf lift was converted to other uses.

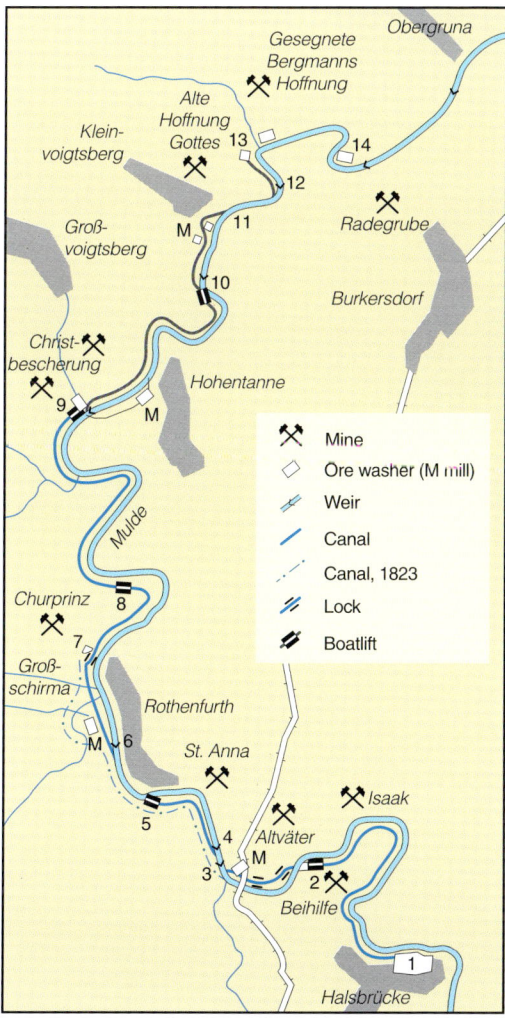

Fig.124 Map of the Halsbrücke canal system
Guide to map
1   Halsbrücke furnace
2   Rothenfurth lift and Isaac ore washer
3   Weir from 1822/3, with new canal shown as dotted line
4   Weir for the ore washer at the Anna mine
5   Lock below the mine
6   Weir for the Großschirmaer mill
7   Mine, stamping mill, warehouse and possible lock at the Churprinz mine
8   Schuman lock
9   Christbescherung ore washer, boat lift and weir for the Hohentanner mill
10 Lift and weir at the Alte Hoffnung Gottes mine
11 Ore washer and double lock at the Alte Hoffnung Gottes mine
12 Weir at the Gesegnete Bermanns Hoffnung mine
13 Ore washer at the Gesegnete Bermanns Hoffnung mine
14 Obergrunaer Forge

(approx. 10.94 m) or 35 Rhineland feet (10.98 metres)[85] onto the higher bank. The upper canal was enclosed by side walls, within which were two rows of dam boards about 5 feet (1.57 metres) apart. The space between was filled with clay, and formed the dam or weir which separated the upper canal from the lower canal. Both side walls were built higher than the upper canal and continued at the same height over the end of the lower canal so that they partly carried the movable hoist for lifting the boats and partly carried a simple roof. On the inside edges of the wall lay beams which were fitted with wooden teeth. Into the latter were engaged

85

the teeth on the wooden wheels carrying the hoist, because here, just as with the earliest attempts to use steam locomotives, the friction of a wheel against the track was considered insufficient to prevent slippage. The hoist, which sits on four such wheels, consisted of two drums around which both ropes to which the boats were attached were wound. Each of these drums was fitted with a gearwheel and these were set in motion by a gear train turned by means of a crank. The previously-mentioned ropes were not however fastened directly to the boat, rather was each fed through pulley blocks, which each had four sheaves. On the lower blocks there were two chains, fitted with hooks which were hooked onto four strong shackles on the edge of the boat. As soon as a boat which was to be raised or lowered was fastened in this way, then four or two men, depending on whether the boat was loaded or empty, turned the crank and raised the

boat sufficiently so that when moved horizontally it would not touch the dam. It hung then freely on the hoist and was moved by this over the other part of the canal into which it was then lowered. The cargo carried was only about 20 Centner (circa 1 ton) per boat.[86]

It took about one hour to pass the lift. The hoisting apparatus was substantially rebuilt after Hagen's visit, after which the tonnage carried was increased to 3 tons per boat. Initially about 250 to 600 Centner (a Centner is 100kg) of ore were carried weekly and over a navigational period of 300 days this results in about 550-1,300 tons per annum. Over the years 1824-1826 a total of 2720 tons were carried and in 1830 about 1100 tons.[87] In the early years, after the opening of the lift, it was possible to take a pleasure trip on the canal and to be 'lifted'. Such trips became rarer around 1804.[88]

Fig. 125 The Halsbrücke boat lift, the world's first vertical boat lift. From a model in the Mining Museum in Freiberg.

In 1830 closing the canal was discussed as the operational and maintenance costs had risen tremendously. However, it was still in service in 1868, when its fate was officially sealed on the 4th July by an order from the Mining Office. This ended the use of canals in the Freiberg mining district.[89]

The structure of the Halsbrücke lift and the dry interchange basin on the former upper canal survive on the banks of the Mulde and were lovingly restored in 1988. An information panel can be found on the roadside close to the river opposite the lift.

Shortly after the completion of the Churprinz Canal, the construction of an extension, the Christbescherung Canal, began on which there is a second structure, the Christbescherunger Hebehaus or Großvoigtsdorf lift,[90] and a third lift was also planned.[91] The Christbescherung Canal remained unfinished and it is quite probable that the Großvoigtsdorf lift was never put in operation. The third lift did not get beyond the planning stage.

Compartment boat hoists

Although not often thought of as a lift, the hoists designed by William Bartholomew for his compartment boat system in 1862 do mark the beginnings of a change in scale in boat lift construction. The system was designed on the Aire & Calder Navigation for loading coal from compartment boats, also called Tom Puddings, into coastal ships with a minimum of handling. The Tom Puddings were loaded with around 40 tons of coal at a colliery and towed down to the port of Goole in trains, usually of twenty-one compartments. There the train was uncoupled and each was floated into a cage, the equivalent of a caisson on a lift, and raised out of the water by a hydraulically-powered hoist. When raised, the compartment and cage were tipped so that the coal emptied down a chute into a waiting sea-going boat.[92]

Originally there was just one hoist, but by the First World War there were five at various points around the docks, including a floating one. The

Fig. 126 No.3 Hoist at Goole during testing around 1900.

system worked until 1986. One of the hoists has been preserved, but unfortunately there is no hydraulic power, so it remains static. Nearby there is a waterway museum with displays and a video to explain the operation of the hoists. An electrically-operated hoist, introduced in 1964 at Ferrybridge Power Station for 170 ton compartment boats, is still in operation.

Lifts in the Netherlands

Electricity was used to power a lift which was built in Amsterdam in 1916 which worked until 1954. A similar lift was built at Brockerhaven and this has been preserved and is described here.

There was an overtoom at Broekerhaven (see page 14) which was replaced, in 1923, by a dry lift without a caisson. The lift was designed to handle 30 boats with a total weight of fifteen tons per hour (up to four boats per lift). It took six months to build and the lift, or overhaal, came into service on the 23rd February, 1923.

It has two (an upper and lower) reinforced concrete basins, which are laid out as an S-shaped dock at two levels. Across both basins are two steel frames, which are connected by two parallel steel lattice girders. The boats are floated over a steel cradle which can be raised and lowered by four steel cables connected to an overhead travelling crane. The crane moves on rails along the inner side of the frames and can move from above one basin to the other. Two electric motors driving through gears operate the overhead crane.[93]

The lift was mainly used by the small flat-bottomed boats which served the numerous local market gardens. Because of changes in transport, particularly after the Second World War, use of the Broekerhaven lift declined and in 1981 it was taken out of service. In 1984 it was designated as a technical monument and was renovated in 1993.[94]

China

There has been a further rebirth of the dry vertical lift in the People's Republic of China. In Hubei province, in 1967, a trial 26 metre lift was built for 20 ton boats at the Puji dam on the Lushui River. It has a lift of 26 metres and the platform is 16.10 metres by 3.80 metres.

A second lift was built at Danjiangkou on the Hanjiang River where a dam was bypassed by a lift and incline (see page 64) for 150 ton boats. On the lift, a rubber floored platform and suitable space between the ropes makes it possible for ships built with a flat bottom to sit on the lifting platform. Gates could be fitted to

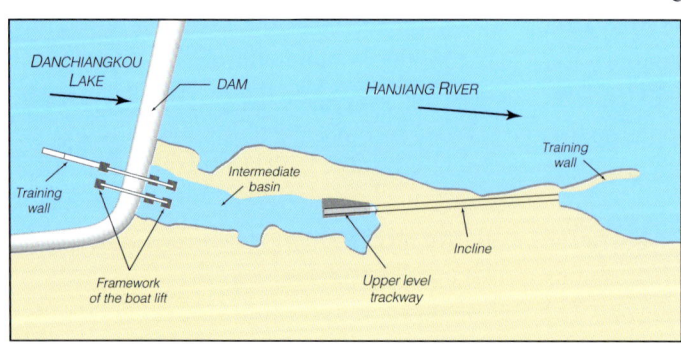

Fig. 129 Layout of the vertical and inclined lift system at Danjiangkou, China.

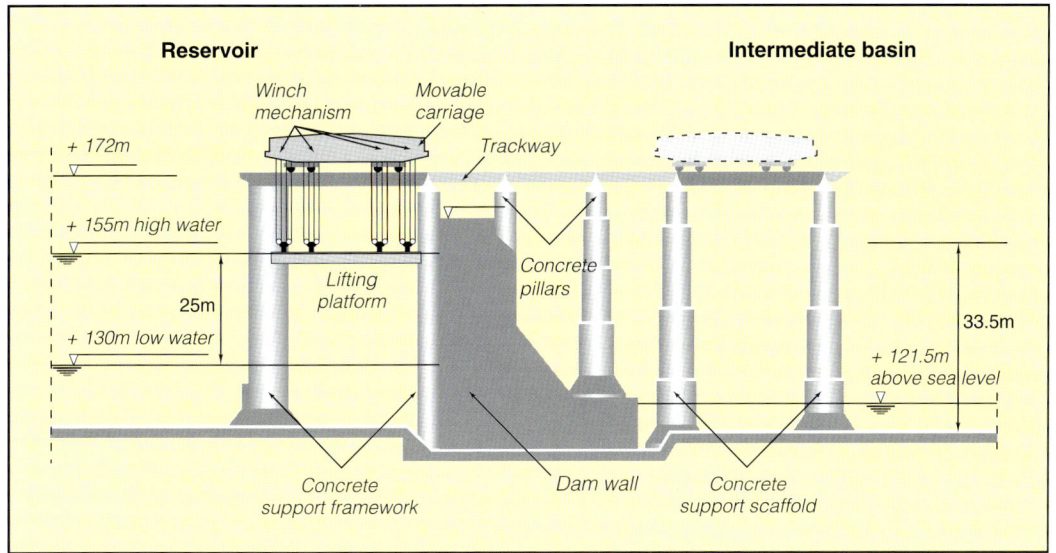

Fig. 130 Cross-section of the lift at Danjiangkou, China.

its ends so that it could become a water-filled caisson, though reducing its effective length. The winches are fitted to a motorised carriage which runs on rails above the dam wall allowing boats to be carried from one side to the other. Raising and lowering takes place between the concrete pillars which also serve as guide scaffolds.

Entering service in 1973, it has a lift of 45 metres. The platform is 32 metres by 10.70 metres and when used as a caisson its length is reduced to 24 metres with a water depth of 0.90 metres. The average lift and lowering speed is 11.2 m/minute.[95]

The final dry vertical lift in China is on the Hanjiang River at Angkang above Danjainkou and has a lift of 87.90 metres. It is designed for 100 ton boats and the platform is 24 metres by seven metres. All three lifts were built alongside hydro-electric dams and were, to some extent, experimental and part of a project for further projects (see page 131).

Diving Locks

The main problem with wet lifts is that the combined weight of water and caisson require both strong supporting cables or chains and substantial lift structures. One way around this difficulty was the diving lock, where the caisson was completely enclosed and submerged in a water-filled shaft. By varying the weight of the caisson slightly, it could be made to rise or fall in the shaft. Much of the technology used for their construction is similar to that for locks, though the need for efficient seals at the ends of the enclosed caisson was always the main problem with this type of lift. There are also safety implications for those on the boats within the caisson.

The Somersetshire Coal Canal: Combe Hay

The diving lock was developed during the late eighteenth century. In 1792 an English patent, no.1892, was given to Robert Weldon, which is described by Hagen as follows:

The caisson is formed by an enclosed water-tight cylinder, which is sunk in a well. The level of the water in the latter should, however, be well above the water level in the upper canal. The caisson must contain enough water for a boat to sail into it, such that with the addition of ballast it weighs precisely the same as the water which it (the caisson) displaces. It can be seen that, so long as it is completely immersed, it will be in balance at any depth and minimal power is required to raise or lower it. This should be achieved by a pressure pump, by means of which the people on the boat can pass a quantity of water from the well into the cylinder, or vice versa. When, in this way, the correct level was reached, the cylinder

89

needed a waterproof connection with the edge of the well in order to prevent the water flowing out of the well and after this was made the large end gate could be opened allowing the boat to sail from the canal into the cylinder or vice versa.

It was a bold invention and it was actually put into operation when the engineer, William Smith, built such a lock on the branch of the Somerset [sic] Canal, at Dunkerton, not far from Bath, between 1796 and 1797. It had a rise of 44 feet (13.41 metres) and could take boats 70 feet (21.33 metres) long and 7 feet (2.13 metres) wide. In the spring of 1798, before the canal was opened, some boats were raised and lowered tentatively. Unfortunately, the walls of the well collapsed and since then neither has this lock been rebuilt, nor has any other with the same arrangement been built.[96]

Hagen's report needs some correction and comment. There seems to have been a trial of the system at Oakengates, in Shropshire, two years after Weldon obtained his patent. This was followed by the so-called 'Caisson Lock' which was also erected under the direction of Robert Weldon at Combe Hay on the Somersetshire Coal Canal. The wooden enclosed caisson had a length of 80 feet (24.38 metres), a width of 10.5 feet (3.20 metres) and a height of $11^{1}/_{2}$ feet (3.51 metres), the rise between the lower and upper canal being 46 feet (14.02 metres). The lock sides were curved, 20 feet (6.10 metres) wide in the middle and $11^{1}/_{2}$ feet (3.51 metres) at either end.

The trials attracted great interest from the public as contemporary newspapers reported; even the Prince of Wales visited the lock in April 1799 and a number of passengers were carried on demonstration trips during that month. A complete passage of the lock lasted on average $6^{1}/_{2}$ minutes.

However, in May 1799, and not in 1798 as suggested by Hagen, the use of this caisson lock came to an end. The main reason for this was unstable masonry, though the principle the diving lock itself had been proved.[97] Its precise location is unknown.

Twentieth century development

More recently, the principle of the diving lock has engaged the interest of several German experts. A variation was suggested for the Oder Spree Canal at Fürstenberg (Eisenhüttenstadt) by the firm Krupp-Gruson from Magdeburg-Buckau using a system devised by Burkhardt.[98] Possibly the most interesting design was that produced by Professor Rothmund in the 1950s and demonstrated for several years on a model which could be used by canoes.[99] A full-sized example has not been built, principally because of obvious safety problems — guaranteeing that the caisson seal is maintained and the danger of the caisson jamming in operation.[100]

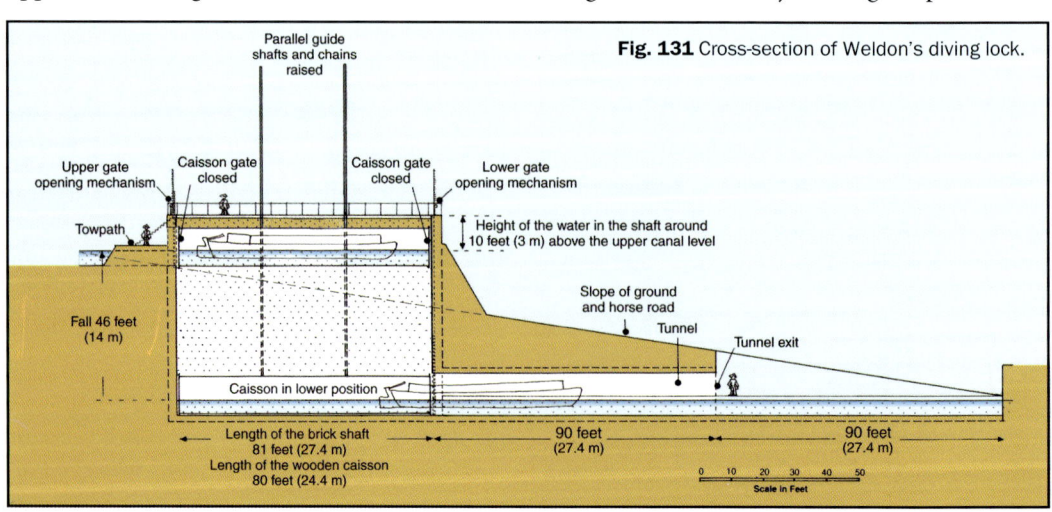

Fig. 131 Cross-section of Weldon's diving lock.

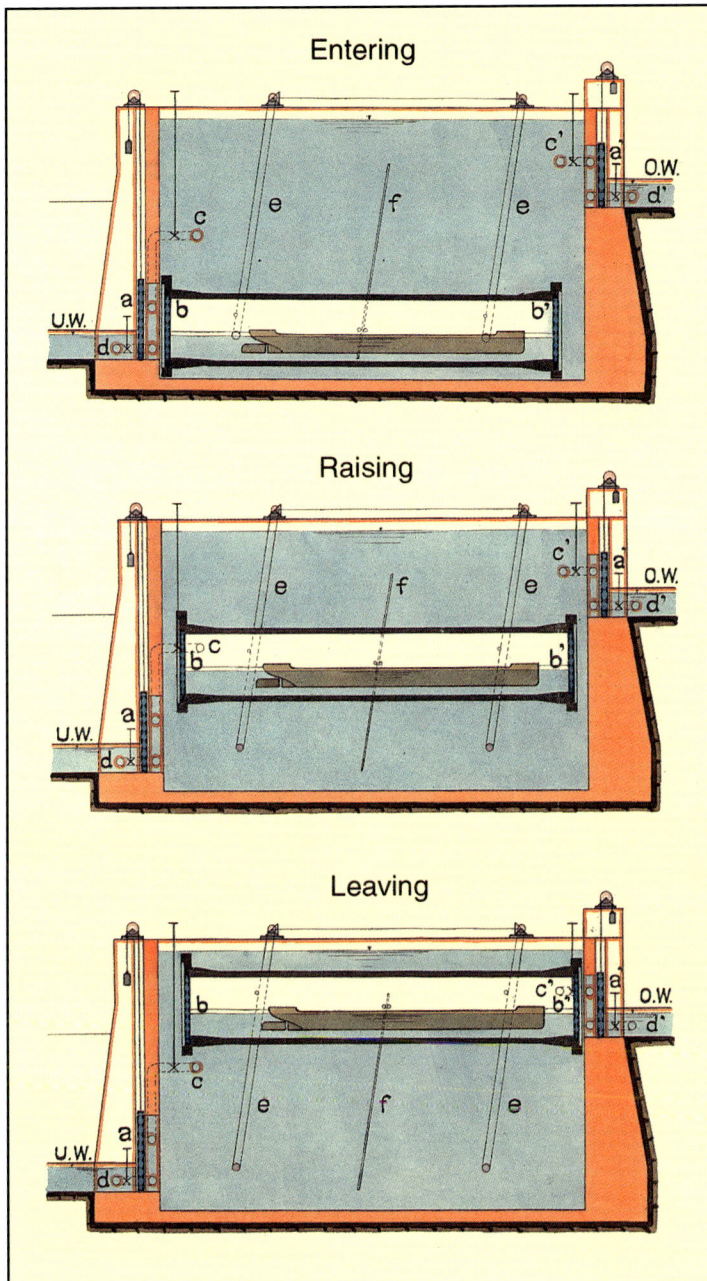

Fig. 132 The diving lock concept as produced by the firm Krupp-Gruson following Burkhardt's system.

Description of operation.
1. Gates a and b are closed after the boat enters the caisson and the space between the two gates is filled with water by valve c. This releases the pressure holding the caisson against the front wall of the shaft.

2. After drive e is set in motion, the caisson rises along track f because of its buoyancy. This has been created by stopping the caisson slightly above the lower canal level, so that when the gates are opened, a small amount of water is released into the lower canal.

3. When the caisson reaches the top, the space between the gates a' and b' is emptied of water by valve d' which results in the caisson being pressed against the end wall. The gates can then be opened and the boat released. The caisson has been stopped such that the water level within is slightly below that in the upper canal. This allows a small amount of water to enter it and thus to create the necessary negative buoyancy for the caisson to descend.

Within the image, the three panels are labelled "Entering", "Raising", and "Leaving".

Flotation lifts
Roland and Pickering's invention

Almost as old as the principle of the diving lock is that of the flotation lift as developed in England which Hagen[101] describes as follows:

At roughly the same time [as the diving lock was invented], namely in 1794, a Patent[102] was granted for a similar invention, also rather impractical but which, however, was built.[103]

The lock chamber [or caisson] consists in this case of a box, which is similar to an ordinary chamber with floor and side walls and open above. It is not itself immersed in water, rather it is carried on flotation tanks by means of light cross-section supports. The water level in the shaft in which the flotation tank which supports the lock chamber moves has to remain below that of the lower canal, because the chamber

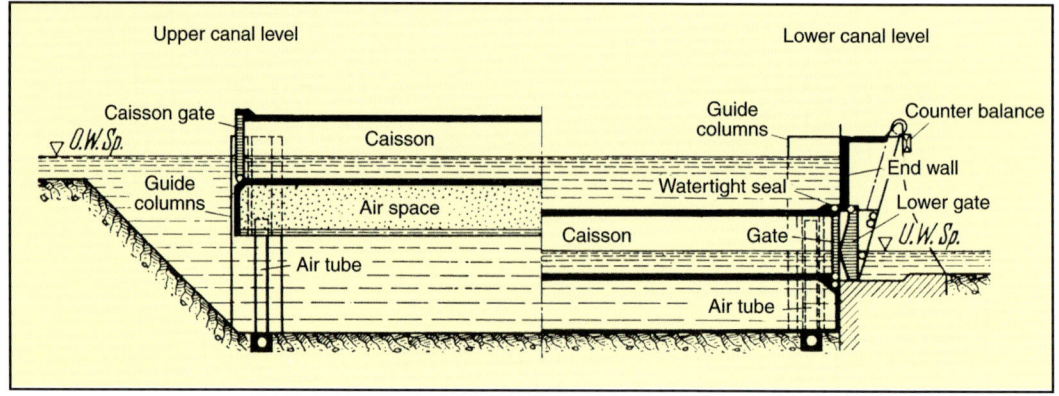

Fig. 133 Cross-section of a diving lock designed by Prof. Rothmund.

could not be immersed without acutely altering the balance. The previously mentioned supports slightly alter the balance if they are immersed in or are out of the water, but their influence is not significant as they have only light cross-section and one can use the pressure caused by them, which is either upward or downward, to set the chamber in motion, or if it is moving, to bring it to a standstill. It is suggested in a report by Chapmann that a lock of this kind was built on the Ellesmere Canal [it was probably at Ruabon[104]]; Dutens[105] looked for it in vain and convinced himself that no such lock existed in England. A very similar arrangement has been recently patented, once more in England, and Simpson[106] has called the invention the hydro-pneumatic elevator.

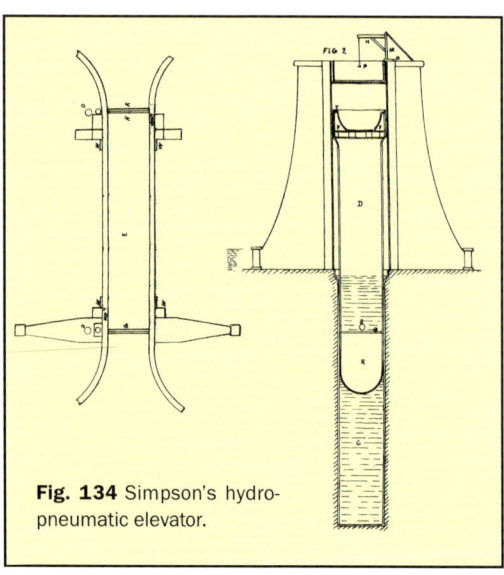

Fig. 134 Simpson's hydro-pneumatic elevator.

Congreve's lock

More like a lock than a lift was the hydro-pneumatic lock designed by Colonel William Congreve and built on the Regents Canal in London in 1815. It was never totally satisfactory and was soon converted to a conventional lock. It worked by raising and lowering the water level in the two chambers of a staircase lock by raising and lowering interconnected air-filled inverted caissons at the bottom of each lock. In effect, air was filling the lock rather than water. It was sealing the caissons and the construction of the lock chambers with a high degree of precision to allow the caissons to rise and fall without leaking which proved too difficult for the technology of the time.[107]

The idea developed

As previously mentioned by Hagen, Mr. Simpson designed a flotation lift around 1850, though the idea was not taken up at the time. Lifts using high-pressure water hydraulics were more successful and several were to be built over the second half of the nineteenth century, including the compartment boat lifts at Goole, the lift at Anderton and similar ones in France and Belgium. It was not until the end of the nineteenth century that the idea of the flotation lift finally was put into practice following numerous theoretical and model investigations. In 1887, in both Belgium and Germany, in the latter by engineer Jebens of Grusonwerk in Magdeburg, lifts were proposed with their caisson supported by submerged air tanks within a well. To raise and lower them, the force was created by adding or subtracting water from the caisson.[108] Further stimulus was given by Prüsmann's suggestion[109] for having several flotation tanks consisting of vertical cylinders in individual wells which were linked together. He patented the various control systems and tested them thoroughly on 1:15 scale models.

Henrichenburg lift

The final proof of the system's practicality was the construction of a flotation lift on the Dortmund Ems Canal at Henrichenburg where a rise of from 14.50 to 16 metres had to be overcome. It was based on a design by Jebens from 1892 for a flotation lift with screw guides (German patent no. 80531).[110]

In 1886, a statute was passed for the building of a waterway from Dortmund, in the east of the Ruhr, to the seaport of Emden. It was to form the first part of a system linking the Rhine, Ems, Weser and Elbe. The Dortmund Ems Canal, built between 1892 and 1899, was built for *Gustav Koenigs* type boats, 67 metres by 8.2 metres and carrying about 700 tons at a draught of 2.00 metres.

Consequently, the caisson at Henrichenburg was designed with length 70 metres, width 8.60 metres and depth 2.50 metres. The weight that was carried by the five flotation tanks was 3,000 tons, of which 1,550 tons were water, 800 tons

Fig. 135 Construction of the first Henrichenburg boat lift.

the weight of the iron caisson and supports and 600 tons for the flotation tanks. The 27.50 metre deep wells in which the cylindrical flotation tanks, each 8.30 metres outer diameter, 12.88 metres high and with a displacement of 620 m³, rise and fall have a minimum diameter of 9.20 metres and are connected by 12 cm wide tubes, so they all have the same water level. The iron supports fitted to the tanks for the caisson are 17.90 metres high and at their top are lattice cross frames, 4.60 metres apart, on which the main supports for the caisson sit.

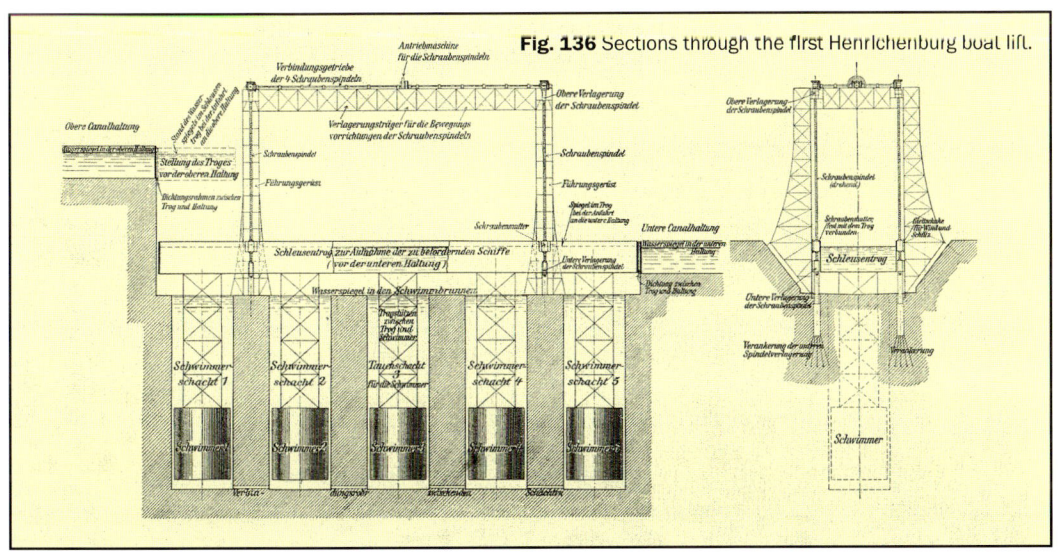

Fig. 136 Sections through the first Henrichenburg boat lift.

Fig. 137 The first Henrichenburg boat lift.

The whole out of balance load, consisting of the flotation tanks, the caisson supports and the caisson itself, were both held level and set in motion by four vertical screwed spindles which could rotate and which were fitted to the main frame. They passed through threaded nuts fitted on extensions of the cross beams forming the transverse supports of the caisson. The spindles were connected by gears and shafts to a single drive. Everything was so well balanced that it was not necessary to add or remove water from the caisson.

The threaded spindles were held by bearing housings at top and bottom so that the caisson could be kept horizontal. The spindles and bearings were large enough so that they could hold the weight of the caisson and flotation tanks in an emergency. The spindles were driven by an electric motor of 150 HP (110 kilowatts), which was located in a small house in the middle of a platform over the upper canal entrance. Both the canal and caisson were protected by lift gates which were balanced by counterweights working over pulleys on a frame over the gate. These frames were also used to solve the problem of sealing the connection between the caisson and the upper canal.

It took about 25 minutes to pass through the lift, of which about 18 minutes was taken by pulling boats in and out of the caisson with

Fig. 138 The synchronisation gears at the top of the lift which link the main drive shafts.

Fig. 139 Opening of the first Henrichenburg boat lift by Kaiser Wilhelm II.

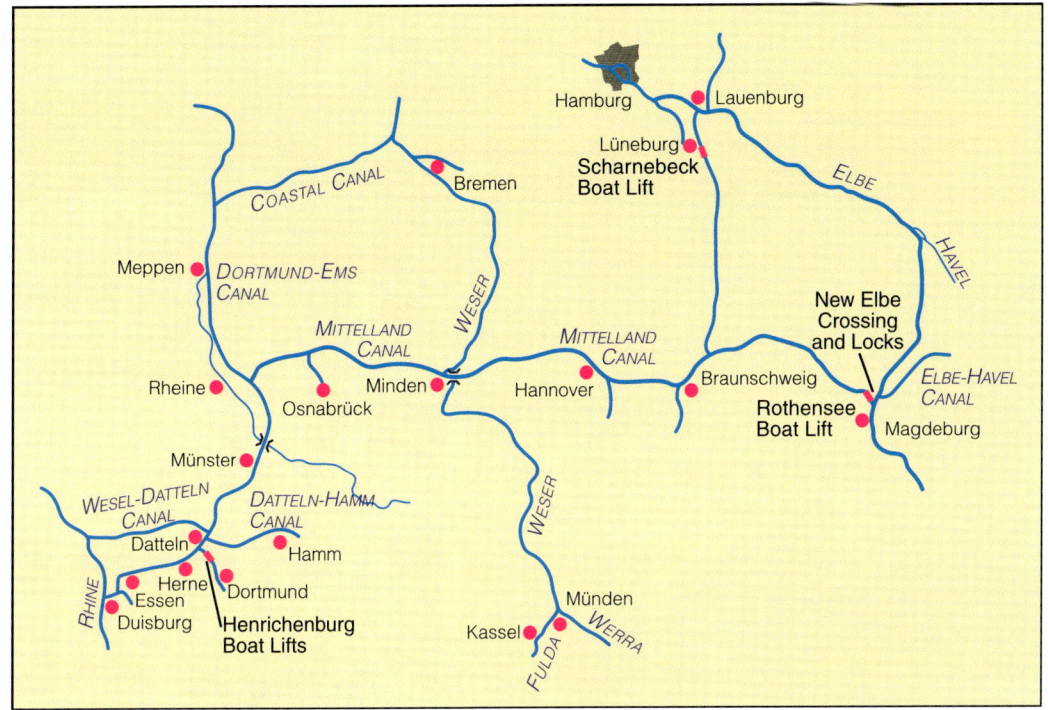

Fig. 140 Location of the Henrichenburg, Scharnebeck and Rothensee boat lifts.

capstans. Its average speed was about 11 cm/s.

Construction began in the summer of 1894 and the caisson was raised for the first time in June 1898. Full trials were undertaken in March and April 1899. The Dortmund Ems Canal and the lift were opened ceremonially on 11th August 1899 by Kaiser Wilhelm II.[111]

Because of increasing traffic, a shaft lock was built near to the lift in 1913. The lock was replaced in 1989 by a larger one.

The old Henrichenburg lift was taken out of service in 1970 after seventy years of operation. A few years earlier, in 1962, a new lift which will be described later had entered service. After many years of argument over the future of the old lift and to the delight of those interested in waterway history, it was declared an industrial monument and so will survive into the future. A waterways museum has been developed around the lift, making it accessible to the public and also preserving various tugs and other boats on site. Throughout the year there are lectures and exhibitions on different aspects of shipping and waterways.

Rothensee lift

A flotation lift similar to Henrichenburg was also built at Rothensee, near Magdeburg, in Germany, as part of the eastern end of the Mittelland Canal. There was to be a double lift at nearby Hohenwarthe and an aqueduct over the Elbe, but these were not completed because of the Second World War.

At Sülfeld, the Mittelland Canal falls around nine metres from its summit level to a height 56 metres above sea level. It was to keep to this level for 85 km before descending to the Elbe Havel Canal after crossing a 950 metre long aqueduct over the Elbe.

About 2.5 km beyond the aqueduct, there was to be the descent at Hohenwarthe, while 1.3 km before it the 5 km long Elbe Abstiegkanal (descent canal) branched off to the south. About 500 metres from the junction is the Rothensee lift, forming part of the Abstiegkanal. The canal created a link to Magdeburg that could be used by traffic to and from the upper Elbe and the Saale. It formed the southern section of the Mittelland Canal. Canal traffic going down the

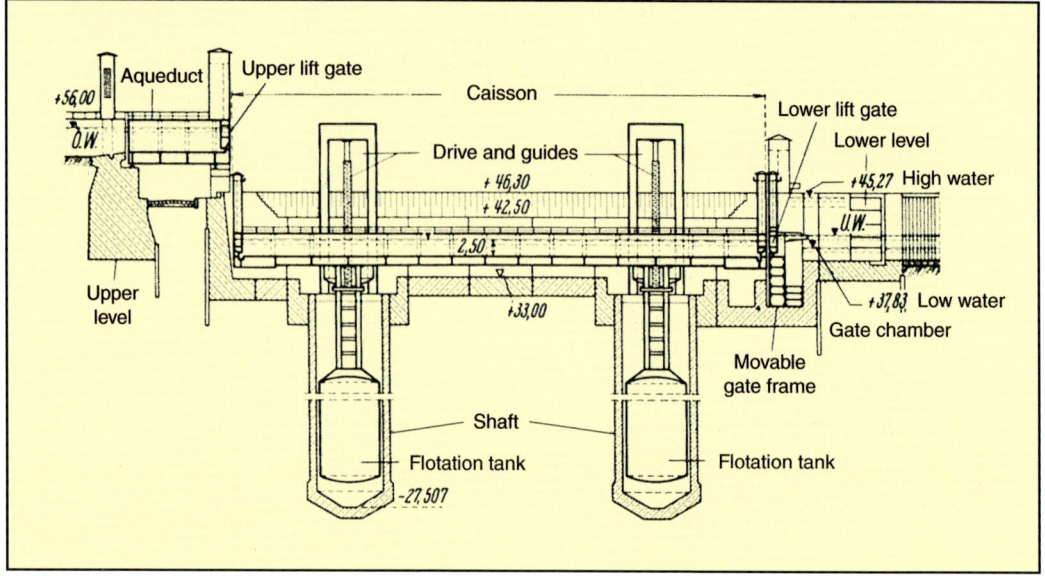

Fig. 141 Cross-section of the Rothensee lift.

Elbe to Hamburg was to use the double lift at Hohenwarthe and then the lock at Niegripp, while traffic between the Ruhr industrial area and Berlin or areas further to the east continued straight on after Hohenwarthe. If there was a problem with one of the lifts, it would have been possible to reach the Elbe by the other.

Fortunately, both lift sites would have had comparable rises (18.67 m at Rothensee, 19.30 m at Hohenwarthe) and so their designs were similar. As much greater traffic was expected at Hohenwarthe, it was planned as a double lift.

Because of the Second World War, only the flotation lift at Rothensee was completed. The double lift at Hohenwarthe did not get further than the initial construction stage and only the end abutments, the river piers and a small part of the aqueduct on the left bank of the Elbe were built.

Recently, as part of 'Projekt 17', a German re-unification transport project, the Elbe aqueduct has been built, completing the Mittellandkanal after a 60-year delay. However, instead of a lift at Hohenwarthe, there are two shaft locks.[112]

Only the Rothensee lift is described here.[113] Its upper entrance is off the high embankment of the Mittelland Canal and it can be closed off by a lift gate which is also used as safety

Fig. 142 The foundations for the Hohenwarthe lifts in 1985.

gate. The lift is 12 metres beyond this and is reached by a short aqueduct. The lift comprises a concrete chamber for the main structure, the caisson, the water-filled shafts including the flotation tanks and the framework on which the caisson is carried.

The bottom level of the lift chamber is deep enough to contain all the moving parts of the lift including the caisson. There is a wall at its lower end so that the caisson can be held against the end of the chamber. In the bottom of the chamber, two shafts were excavated into which the flotation tanks fit. They are completely filled with water and made deep enough so that when the flotation tanks are in their highest position they are still fully covered by water and thus always exert the same lifting force. The original design had four flotation tanks of 10 metres diameter but the type of subsoil allowed the flotation shafts to be made deep enough such that only two tanks were needed. This reduced the building costs by 4%. Their construction required the ground to be frozen as the subsoil was water-bearing up to 40 metres deep and there was also greywacke (similar to sandstone) with water-filled fissures below. Calculations showed that with the construction of a closed sheet pile dam followed by a lowering of the groundwater, it would be possible to excavate up to 70 metres and still curtail the water influx. Without this, construction would have taken much longer. To excavate the shafts, the ground was frozen for about four months to allow the excavation of the 13 metres diameter pits. Two circles of 16 and 19 metres diameter, each with thirty refrigeration pipes, were installed. The inner pipes reached just one metre deep into the greywacke, while the outer ones went five

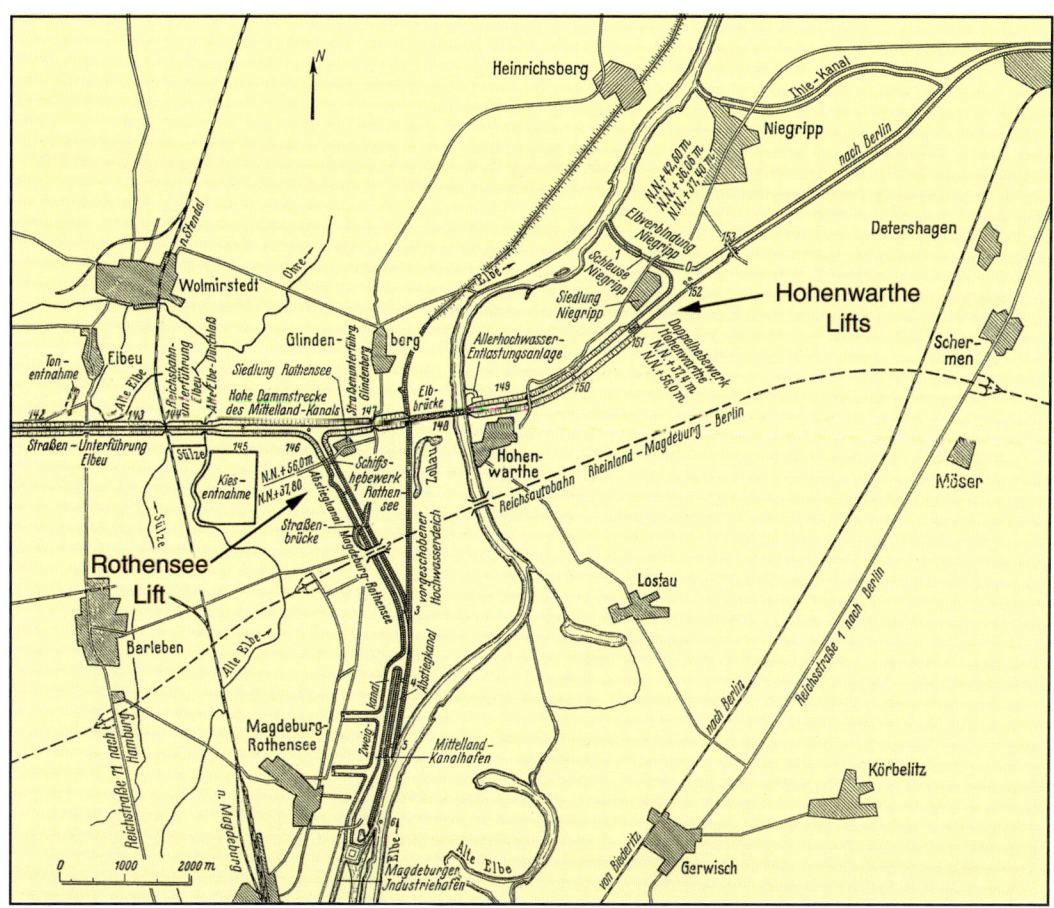

Fig. 143 A 1930s map of the Elbe crossing and the Rothensee and Hohenwarthe boat lifts.

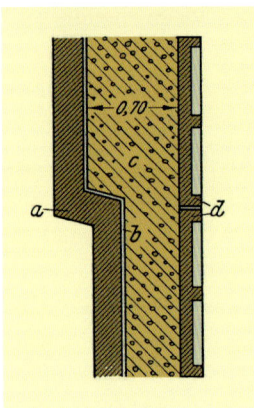

metres below the bottom of the excavation. Freezing conditions were created by two ammonia compressors of 350,000 kcal/h and to protect the frost wall from damage the shafts were sunk by hand with pick and shovel.

The side walls of the flotation shafts are formed by 3.5 cm thick cast-iron tubbing, whose joints were sealed with a 2 mm thick lead gasket and then covered with a 50-70 cm coat of concrete made from 400 kg high-grade Portland cement. To keep the groundwater out, this concrete wall was covered with a 1 to 2 cm layer of heavy bitumen, sprayed at 170° C and 90 lbs/in² pressure before the wall was covered with 20 cm of high alumina concrete. Tests had shown that this concrete would prove highly suitable. The increase in temperature of the high-alumina cement as it sets and its speedy initial hardening made it possible to master these extreme conditions. The bottom of both shafts were constructed in the same way as the shaft lining. The flotation tanks are about 36

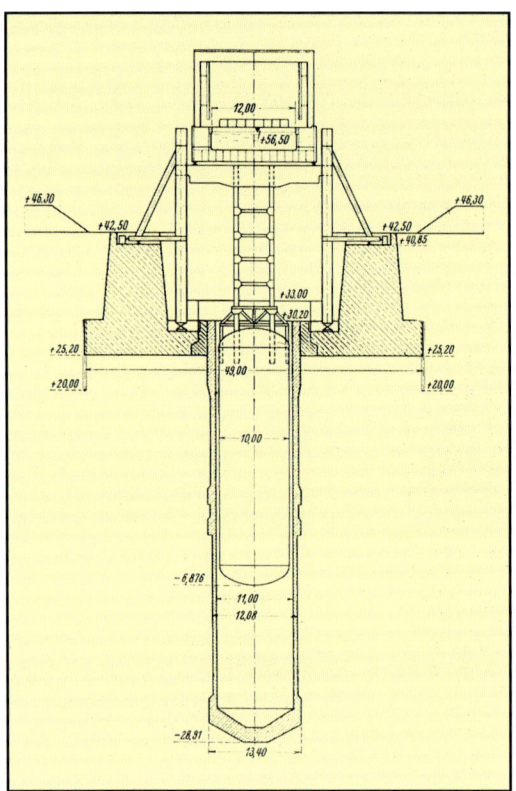

Fig. 146 Cross-section through the Rothensee lift.

metres high and a diameter of 10 metres. Their buoyancy is equivalent to 2 x 2700 tons or the weight of the movable section, about 5400 tons, so that only eight small 44 kilowatt motors are required. They can operate the system with a variation in weight of 80 tons, corresponding to a variation of water level of ±7 cm in the caisson. The forces are fully balanced if the caisson has a nominal water depth of 2.50 metres in its upper position. As it starts to descend, an imbalance is created because of the extra water which is displaced by the caisson supports. To even this out as much as possible, on the central axis of each flotation tank there is a closed tube filled with air compressed by water pressure. As the lift descends, the column of air is compressed such that the additional displacement caused by the caisson supports entering the water is equalised by a decrease in the volume of air. The flotation tanks are welded from St 37 steel and support the caisson on a 22.5 metre high framework.

Fig. 145 The top of one of the flotation shafts.

Fig. 147 One of the four threaded spindles and drive nuts which raise and lower the lift.

The caisson is 85 metres long, 12 metres wide and 2.50 metres deep. Its is supported by the caisson bridge, two plate girders about two metres high and at either end of the caisson there are lift gates. Four frames are fixed to the floor and walls of the caisson chamber and each carries a vertical stationary screwed spindle. These four spindles are masterpieces. Each is machined from Siemens-Martin steel 27.30 metres long with an outer diameter of 420 mm.

The blanks were shaped by forging and then straightened after heat treatment. Machining was on a specially constructed lathe and a 165 mm diameter hole was bored through them so that any flaws could be found. To avoid damage during transport, and erection and installation of the caisson guide frames, the spindles were encased in special lattice-work frames.

Each spindle is fitted with a screwed nut and these are turned by drive motors to make the caisson move. The initial caisson speed is 0.02 m/s, rising to a maximum of 0.15 m/s before returning to 0.02 m/s. On average, it about three minutes to rise or fall.

To keep weight to a minimum, the movable parts of the lift were made from St 52 steel apart from the flotation tanks. The steel girder construction, again with the exception of the flotation tanks, is riveted. The upper canal is

Fig. 148 Rothensee lift undergoing maintenance. The caisson and forebay have been drained.

Fig. 149 A view from the end of the caisson.

protected by a normal lift gate, but for closing off the lower entrance a special gate had to be designed to accommodate the great variation in water levels on the Elbe. A normal lift gate would have had to be over 10 metres high to allow for these variations. To overcome this problem, a 'Schildschütz' or gate with adjustable height was built. Fitted at the bottom entrance to the lift chamber, the Schildschütz hangs in a framework which can be raised or lowered. It is sealed against the down-stream end of the

Fig. 151 The Schildschütz and guide frames are at the lower end of the caisson.

chamber by the pressure of water in the lower canal. In the Schildschütz, there is a 12 metre broad opening for boats in its upper side. This is closed by a lift gate which hangs on a frame fitted to the Schildschütz. The whole structure can be raised in stages (six stages of 1.05 metres and one of 1.14 metres), so that for any water level in the river there is enough depth of water in the gateway for boats. At low river levels, the Schildschütz is lowered into a pit in the entrance chamber. To restrict the depth of the pit, the Schildschütz was divided into a top section, including the lift gate, and a lower section. The lower part can be detached from the upper and moved horizontally in the Schildschütz chamber. The upper part can then be lowered alongside it. To allow this to take place, there is a mitre gate in the lower entrance channel. After it has been closed, the water is pumped out of the chamber and the work can be done in the dry.

To finish, some statistics about construction and operation of the lift. Around 225,000 m³ of earth had to be moved, of which about 15,000 m³ were material excavated from the shafts. Some 55,000 m³ of concrete and 10,000 tons of steel were

Fig. 150 The operating gate fitted within the Schildschütz.

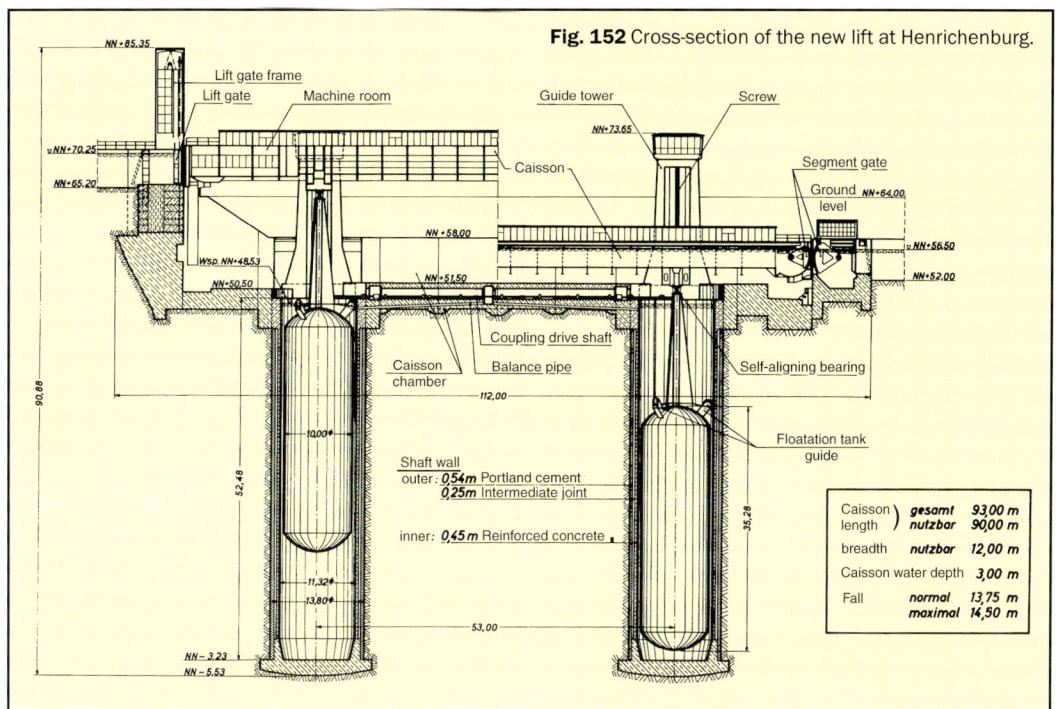

Fig. 152 Cross-section of the new lift at Henrichenburg.

NN + 85.35
Lift gate frame
Lift gate   Machine room
Guide tower
Screw
NN + 73.65
NN + 70.25
NN + 65.20
Caisson
Segment gate
Ground NN + 64.00 level
NN + 58.00
▽NN + 56.50
Wsp. NN + 48.53
NN + 50.50   NN + 51.50
NN + 52.00
Coupling drive shaft
Caisson chamber   Balance pipe
Self-aligning bearing
112.00
90.68
10.00⌀
52.48
Floatation tank guide
Shaft wall
outer: 0.54m Portland cement
0.25m Intermediate joint
inner: 0.45 m Reinforced concrete
35.28
11.32⌀
13.80⌀
53.00
NN – 3.23
NN – 5.53

Caisson length	gesamt	93,00 m
	nutzbar	90,00 m
breadth	nutzbar	12,00 m
Caisson water depth		3,00 m
Fall	normal	13,75 m
	maximal	14,50 m

used. To work the lift, six operators per shift are required, and there is also an electrician and four skilled maintenance men. On one shift, there is also a plant manager.

The lift came into service on the 30th October 1938 after six-years under construction and has given complete satisfaction. Alongside the lift, a shaft lock has been built as part of 'Projekt 17' to enlarge the Mittelland Canal to 1,350 ton standard suggesting that today shaft locks are more economic than lifts for rises of 20 to 30 metres.

The second Henrichenburg lift

On the 31st August 1962, the most modern flotation lift entered service on the Dortmund Ems Canal at Henrichenburg/Waltrop. It is similar to Rothensee and, unlike the older lift at Henrichenburg, has two flotation tanks. These are 10 metres in diameter, 35.28 metres high and are divided into two cells. Both cells are filled with air which is compressed by water pressure. The pressure of the upper cell is 3.0 bar; the lower cell 4.0 bar.

The caisson has a usable length of 90.0 metres (total length 93.0 m), a width of 12 metres and a water depth of 3.00 metres. The caisson rests on a longitudinal fabricated beam, which is supported by two 20 metre long transverse beams 53.0 m apart, which project beyond the sides of the caisson and also support the caisson side walls. The two

Fig. 153 The upper gate at Henrichenburg.

101

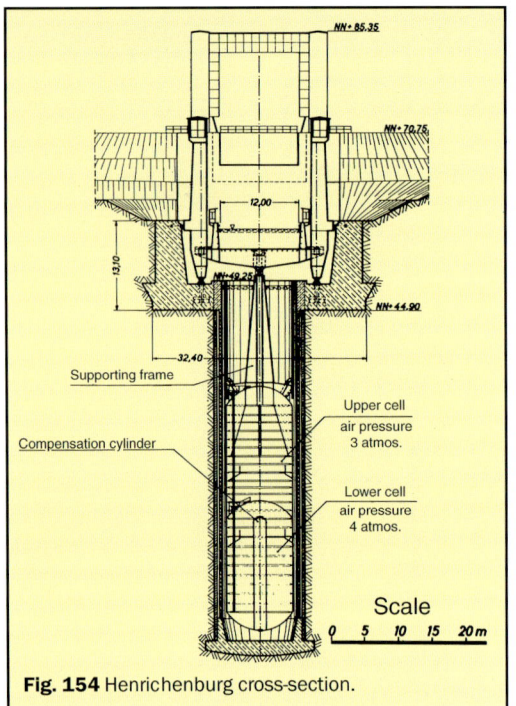

NN+ 85,35
NN+ 70,75
12,00
13,10
NN+49,25
NN+ 44,90
32,40

Supporting frame

Compensation cylinder

Upper cell
air pressure
3 atmos.

Lower cell
air pressure
4 atmos.

Scale

0   5   10   15   20 m

Fig. 154 Henrichenburg cross-section.

main transverse beams are over the flotation tanks to which they are connected by frames.

The total weight of the moving parts is about 5,000 tons; the caisson including the framework about 800 tons, flotation tank 630 tons, water 3500 tons. This is roughly equal to the lift of both flotation tanks, so that to move the caisson only four 110 kilowatt electric motors are required. The caisson is moved by four vertical threaded drive shafts about 20 metres long and outside diameter 372 mm,

which are fitted to guide frames either side of the caisson, the drive motors being at their upper end. Screwed onto them are threaded nuts fixed to the main transverse beams. The shaft drives are connected by horizontal shafts and gears so that they are synchronised.

Sideways movement of the caisson, both longitudinal and transverse, is controlled by guide rollers, fitted to the two main transverse beams, which run on guide rails on the guide frames.

In case of an accident, such as the caisson emptying, overfilling, or the flotation tanks loosing buoyancy, the vertical forces on each spindle are transferred by the guide frames to the foundations. Mechanical double-shoed brakes can, in such circumstances, stop the spindles within 10 seconds.

Segment gates are fitted to the ends of the caisson. They rotate on a horizontal axis and when the caisson end is open lie in a gap under the caisson floor. Segment gates were used for structural reasons, to minimise the height of the lift which would have been greater if lifting gates were used. A similar gate was also used for the lower canal, but since the lift is in an area of mining subsidence, a lift gate was used on the upper canal as this can cope better with any possible variation in the water level.

The average caisson speed is 13 cm/s, so the 13.50 metre rise takes about two minutes,

Fig. 155 An end view of the caisson of the new Henrichenburg lift whilst undergoing repair and showing the segment gate.

including acceleration and deceleration. The time for passing through the lift, including entering and leaving, is about 34 minutes. Theoretically, when operating on a two-shift system for 313 days per year, which allows for holidays, etc., the annual tonnage which can pass through the lift is approximately 15.1 million tons.[114]

Hydraulically-operated lifts

Until the mid-nineteenth century, lifts were comparatively small. The power requirements to overcome bearing friction, etc., were excessive and could only be reduced by improvements to materials and lubrication. The high pressure water-hydraulic lifts, introduced in the second half of the nineteenth century, were the first solution to this problem. In the fifty years after 1875, eight hydraulic lifts were built, excluding the five compartment boat hoists at Goole. (see page 87)

The Goole hoists worked independently, but, until recently, all other hydraulically-operated lifts have worked as twin lifts, though usually with the possibility of independent operation in the event of an accident. Both caissons are fitted onto pistons, whose cylinders are interconnected by a pipe with a cut-off valve V (see Fig 156) and the weight is transferred through high pressure water piping. With the opening of the connecting pipe, the lift works like a large hydraulic balance. If one caisson is slightly overfilled with water, the whole system is put into motion. Since only the friction of the watertight glands, pipes and guides needs to be overcome, merely a small amount of additional water is required. This is particularly important, as any water added during each movement must be drained from the caisson on reaching the lower canal and so is lost from the upper canal. Both caissons must be regulated such that when the water level in the upper caisson has been raised slightly, the lower caisson is raised an equal amount above the level of the lower canal. Movement of the caissons is then controlled by opening or closing the valve on

the pipe between the two cylinders. Water losses caused by poor piston seals have to be replaced by a pump. Apart from altering the water level in the caissons, this is the only energy required to move the caissons.[115]

Anderton lift

The first twin hydraulic lift was built in England in 1874/75 by Edward Leader Williams and John Watt Sandeman following the patent of Edwin Clark.[116]

The Anderton lift is near Northwich and connects the Trent & Mersey Canal to the River Weaver Navigation. The Trent & Mersey Canal runs from the River Trent at Derwent Mouth through Derbyshire, Staffordshire and Cheshire to join the Bridgewater Canal at Preston Brook. It was an important early waterway linking Liverpool and Hull to the Potteries. Originally proposed as a wide canal, during construction it was decided to make it a narrow canal to reduce costs. The section between Anderton,

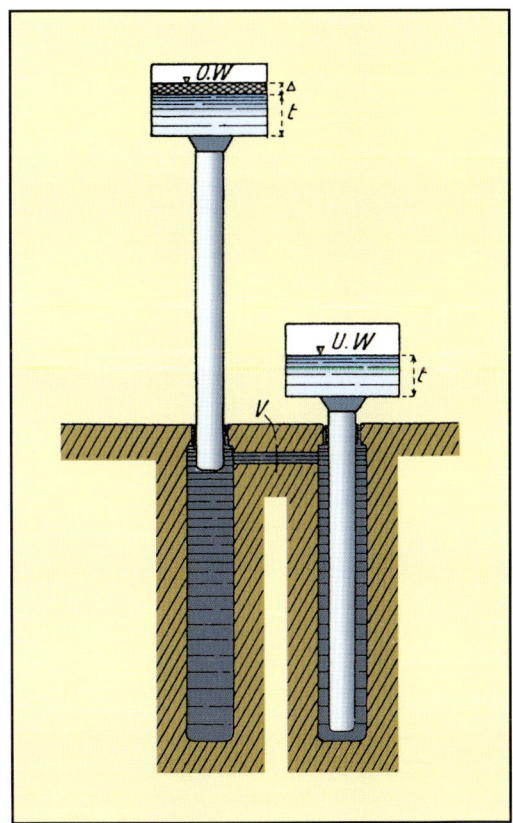

Fig. 156 The principle of hydraulic lift operation.

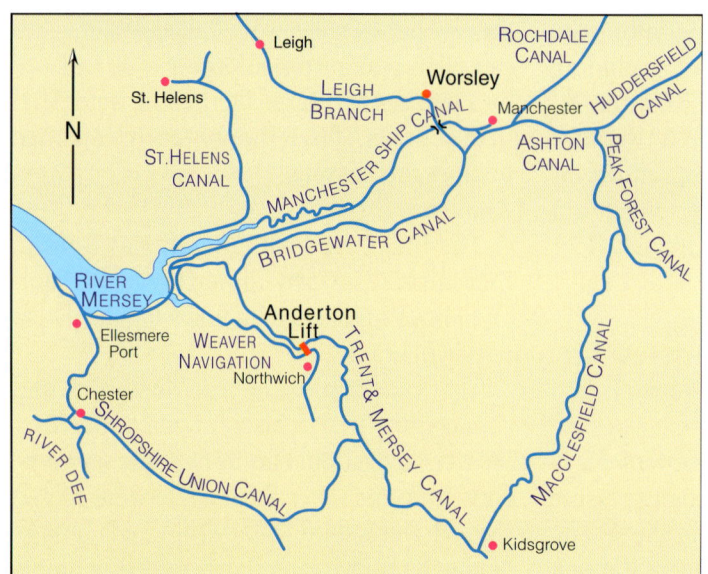

Fig. 157 The location of Anderton lift and the incline at Worsley.

Preston Brook and Middlewich is wide and this is why the lift was designed for wide boats. Completed in 1777, its main line is 93$^1/_2$ miles (150.44 km) long and has 5 tunnels, 76 locks, 164 aqueducts and 109 road bridges. The summit level at Harecastle is 409 feet (124.7 metres) above sea level.[107]

The Weaver Navigation is 17 miles (27 km) in length and originally ran from Winsford to Frodsham and the River Mersey. It now has five locks and weirs and, at the time the lift was built, it could be used by boats of 300 tons capacity. For much of its length, the Trent & Mersey Canal runs almost parallel to it. Northwich is the centre for the Cheshire salt industry as well as having numerous chemical works. At Anderton, the canal and river are virtually alongside each other and it was an obvious location for a link

Fig. 158 Anderton lift around 1900, with salt chutes on either side and the boiler house chimney on the right.

Fig. 159 The original lift at Anderton in 1878.

between them. The 50.33 feet (15.34 metres) rise was bridged when the lift came into service on the 26th July 1875.

It was erected on an island, now connected to the river bank, in the Weaver and canal boats reached the lift across an aqueduct.[118] This was made from wrought iron in three sections of length 30 feet (9.14 metres), 75 feet (22.86 metres) and 57½ feet (17.53 metres) and comprised two troughs, each 17 feet (5.23 metres) broad with a maximum water depth of 5 feet (1.60 metres). The weight of the aqueduct, including water, is 1,050 tons.

The caissons had guides at all four corners and sat on cast iron cylinders three feet (0.914 metres) in diameter. Each of the 75 feet (22.86 metres) long and 15½ feet (4.72 metres) broad caissons could hold either one barge (80-100 tons) or two narrow boats (25-35 tons). The total weight of a caisson and piston was 252 tons. The pistons were also cast iron and operated at a pressure of 550-670 lbs/in² (37-45 bar).[119] In contrast to all later hydraulic lifts, as originally designed the caissons at Anderton entered the water at river level, the

lift's lower chamber being filled with water. Consequently, no gates were needed there, but both the caissons and the ends of the aqueduct had wrought iron lift gates. To form a water-tight connection, there was a bevelled rubber seal which was compressed as the caisson reached its upper position. At the upper level there was a siphon which was used to add the extra 6 inches (15 cm) of water needed by the caisson to make it descend. There was no balancing of the variable forces on the pistons. For this reason and because the caisson had buoyancy when lowered into the water, a 10HP steam pump and an hydraulic accumulator was used to overcome any resistance. The caisson took 2½ minutes to ascend or descend. It was possible to operate the caissons independently using the steam pump, but then the time taken would have increased to half an hour.

The lift was revolutionary and, as such, teething troubles could be expected. It worked almost without problem until 1882, but on the 26th April of that year there was a serious incident.[120] A caisson was at its uppermost position when, just as the additional weight had been added, it began to sink rapidly. An upper shoulder of the groove in the cylinder for the hydraulic seal was found to have broken off and the seal had sprung out. For safety, the Weaver Navigation's engineer, Lionel B. Wells, immediately subjected the other cylinder to a pressure test in case it was damaged as well.

A year earlier, in 1881, it had been noticed that the piston and other moving parts in contact with the water of the Weaver and the Trent & Mersey Canal had begun to wear and show heavy corrosion. It was attributed to the pollution of the water caused by the local chemical and salt industries. The damage was repaired, but there continued to be problems with the cylinders and distilled water was used as the working fluid to help prevent corrosion.

In 1901 electric lights were installed and the following year, because the steam plant was worn out, two 30 HP electric pumps were installed. The gates were also converted to electric operation. In 1904 the Navigation's Engineer, J. A. Saner, produced a report on the condition of the hydraulic plant and lift in which he proposed strengthening the structure and converting its operation to a counter-weighted system.[121] A general repair, with a complete closure of the lift for a long time, would have seriously interfered with traffic, so ingenious measures were decided upon. These are discussed later on page 119.

Les Fontinettes lift

In the summer of 1888, two more hydraulic lifts came into service, in France and in Belgium.

The first was on the Canal de Neuffossé near Saint Omer in northern France. This canal, which opened in the mid-eighteenth century, is now part of the important waterway between Dunkirk (Dunkerque) and the Schelde.

The five locks at Les Fontinettes, with a rise of 13.13 metres, were replaced by a lift designed for boats of 300 tons capacity, the total weight to be raised including the caisson being 800 tons. The piston diameter was 2 metres, while

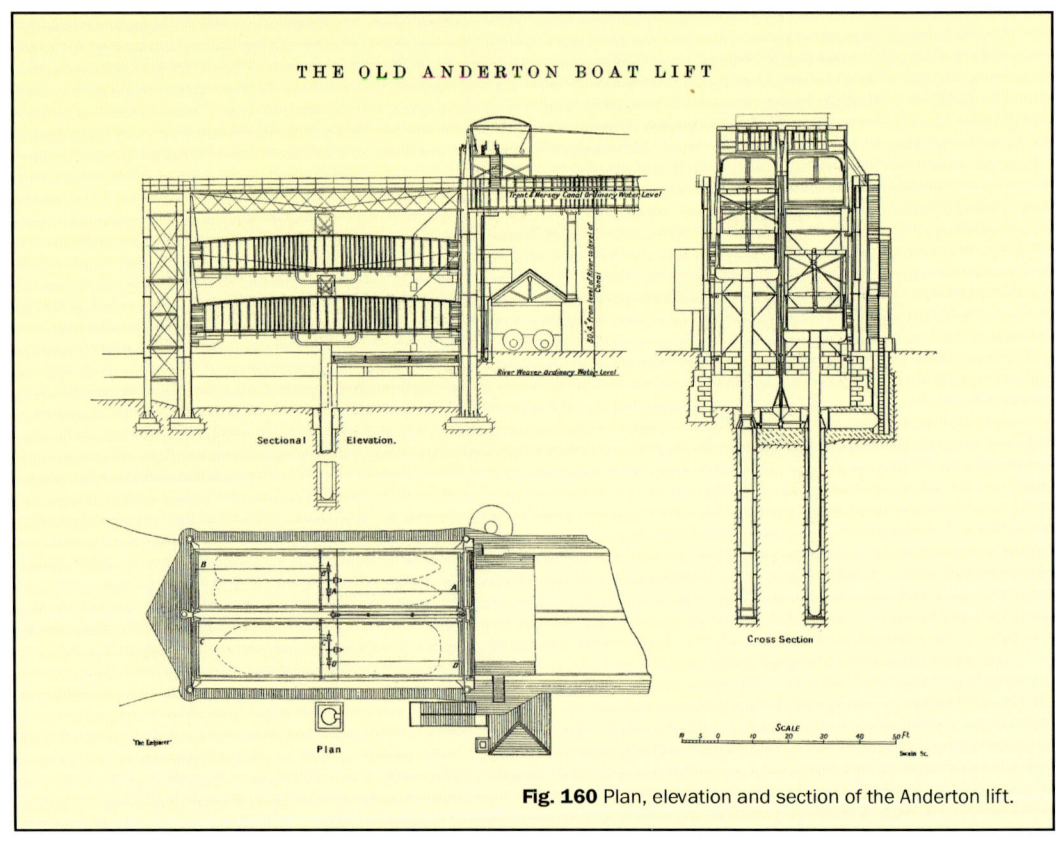

THE OLD ANDERTON BOAT LIFT

Fig. 160 Plan, elevation and section of the Anderton lift.

Fig. 161 Cross-section of Les Fontinettes lift.

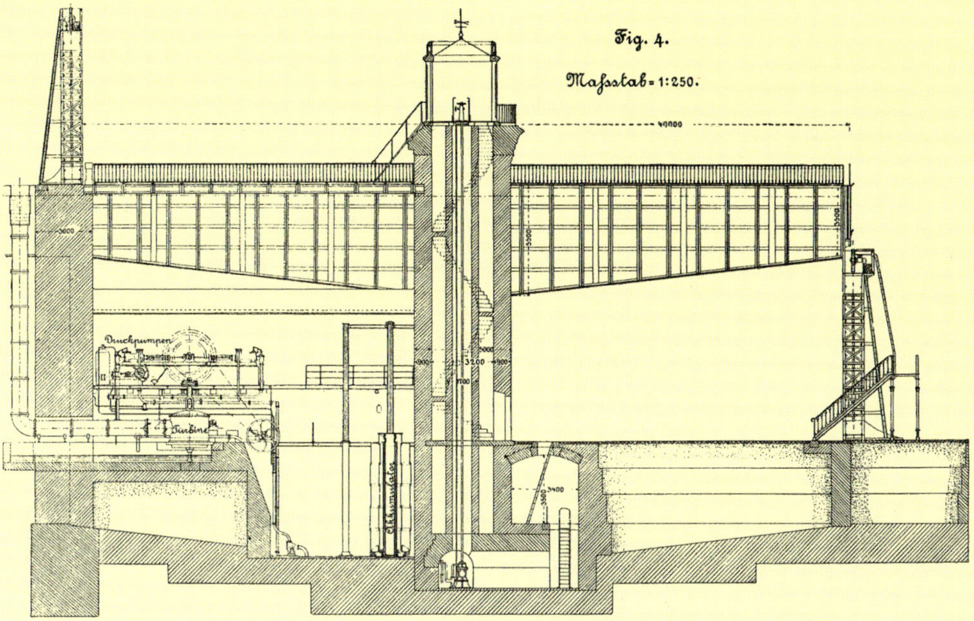

Fig. 4.

Maßstab = 1:250.

the cylinder pressure was 25 atm. (25 bar)[122] The caissons were 40 metres long and 5.80 metres broad with a minimum water depth of 2.10 metres. Under test on 28th April 1888, 43 boats, including 40 carrying some 6,700 tons, were passed up or down in 18 hours. Since no capstans were originally provided, it took about 42 minutes for two boats to pass.[123]

The original scheme was by the Anderton designer, Edwin Clark. There were important alterations during construction, these being carried out under the direction of chief engineer Bertin and assistant engineer Barbet by the Parisian firm Cail et Cie and crucial to the changes was the accident at Anderton in the summer of 1882.[124]

The official opening of the lift was on 8th July 1888. Unlike Anderton, the caissons descended into a dry pit, so there was no weight decrease caused by buoyancy of the caissons. This made it necessary to fit the lower canal with lift gates as well as the caissons and the upper canal.

The variation in load as the caissons rose or fell were fully hydrostatically balanced by a

Fig. 162 A caisson chamber and piston at Les Fontinettes.

Fig. 163 A recent view of Les Fontinettes.

system invented by Clark and Standfield. In the two outer guide towers there were two cylinders (C1 and C2, see fig.165), which had the same cross-section as the main cylinders and which were connected by articulated pipes (G1 and G2) with their respective caissons. When the caisson descended to its lowest position the decreased load on the piston was equalised by passing water from C to A. In the opposite direction when the caisson was in the highest position, the increase in force on the piston being equalised by passing water from A to C.[125] This ingeneous arrangement was not used and the additional weight of 63.5 tons required by the descending caisson was created by adding 30 cm of water from the upper canal.

Sideways movement of the caissons was prevented by guide towers the same height as the pistons and to avoid rotation by wind pressure there was a massive rear wall. Instead of cast iron components, steel was used almost exclusively.

Fig. 164 The aqueducts and rear wall of the Les Fontinettes lift.

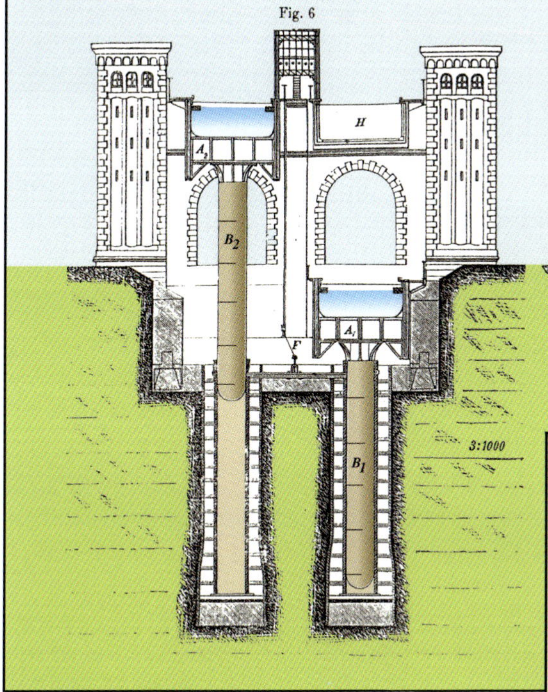

Fig. 165 The hydrostatic load balancing system at Les Fontinettes.

Fig. 6

Fig. 7

However, there were errors in design. The foundations and load-bearing bases of both cylinders were too small. The result was that the pressure on the foundations was too high and they began to sink. After six years, this was bad enough for major repairs to be necessary. By then the lift had made about 25,000 operations, raising or lowering some 41,000 boats carrying 5.5 million tons.[126]

During repair works the entire foundations had to be rebuilt using ground-freezing and this required a two-year closure of the lift. During this time, boats used the original five lock flight.

In 1959 it was decided to rebuild the canal to 3,000 ton standard and the lift was by-passed because it could not be enlarged. In August 1967, it was taken out of service and now boats use a new adjacent shaft lock.[127] Since then the lift has been restored, though not to working order, and there is a small museum alongside. It is easily found by following the signs on the road from Saint Omer to Arques.

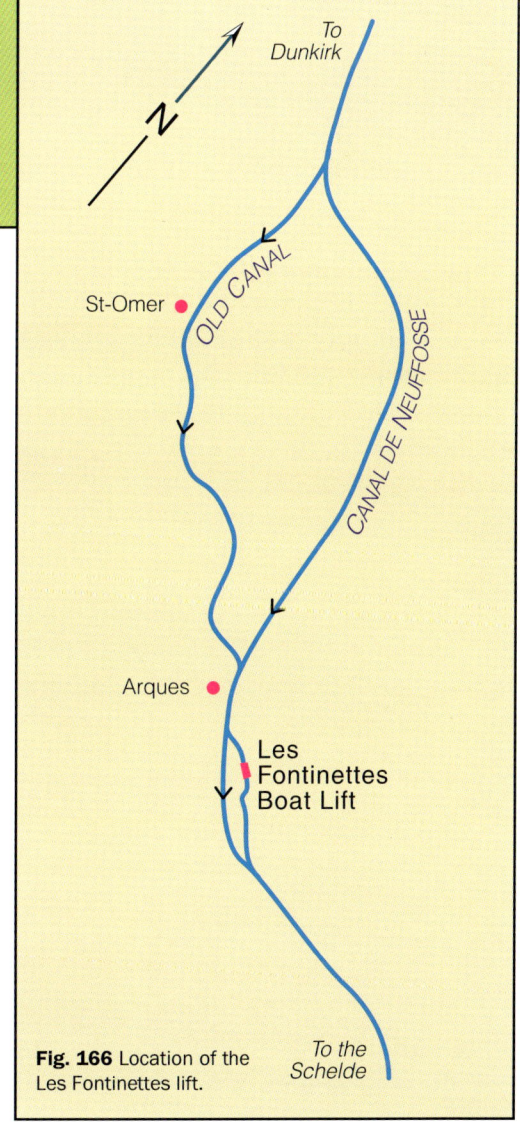

Fig. 166 Location of the Les Fontinettes lift.

Fig. 167 The water turbine used to drive the hydraulic accumulator pumps at Les Fontinettes.

The Canal du Centre lifts

Also in the summer of 1888, the first of four lifts came into service on an 8,600 metre long section of the Belgian Canal du Centre between Thieu and Houdeng-Goegnies. The canal links the Canal de Charleroi à Bruxelles and its surrounding industrial areas with the north French industrial region. Construction of the was begun in 1884 and finally completed for 300-ton boats in 1917. There was a 90 metre rise between Mons and Houdeng and hydraulic lifts were used to overcome 66 metres of this.

Of the four lifts, Houdeng-Goegnies, often called La Louvière, was completed in 1888; Houdeng-Aimeries, Bracquegnies and Thieu were all opened in 1917. At Houdeng-Goegnies the rise is 15.40 metres and the other three 16.93 metres. They all use the same basic hydraulic system, though the earliest lift was somewhat different in structural design. All the caissons were large enough for boats of up to 400 tons and were 43-45 metres long by 5.8 metres broad with a minimum water-depth of 2.4 metres. They sit on 19.44 metre high pistons of two metres diameter. The cylinders, which are made from two metre long cast iron rings of 2.06 metre inside diameter with 100 mm wall thickness, can stand a pressure of 34 atm. (34 bar). This supports a load of 1,048 tons at Houdeng-Goegnies for a descending caisson.[128]

The high-pressure water supply is generated in three power houses alongside the lifts. In each there are two Pelton wheels with horizontal shafts and water is taken from their respective

Fig. 168 The Bracquegnies lift and power house.

upper canals. The power house at No. 3 lift also supplies No. 2 lift, taking its water from the canal above the latter. The two lifts are only 384 metres apart. About 46 m³ of water are need to power one lift operation and about 80 m³ of water are required to create the 32 cm excess depth of water for the descending caisson. If one adds to this the 8 m³ of leakage from both pistons, 7.5 m³ of drive water for the Pelton wheel pumps and 10.0 m³ for the capstans as well as about 2.0 m³ of other water losses, the volume of water needed for one lift operation is about 153.5 m³. Compare this with a shaft lock of equal lift which would need about 4,420 m³ of water.[129]

The caisson guides at Houdeng-Goegnies were, as at Anderton, fitted at the four corners, though there was also a central guide tower. The three later lifts only had guides on the central tower and at the upper canal end of the caisson. In contrast to Les Fontinettes, the support structure was steel latticework and the cast iron cylinders were strengthened by steel bands.[130] Unlike Les Fontinettes and Anderton, there was no aqueduct from the upper canal as this joined the lift at the top of a walled embankment. A road ran behind this wall at lifts 1 and 4.

Lift gates are provided for both the canal and caisson and they are also used to equalise any difference in water levels. When the caisson has reached its upper or lower level, the space

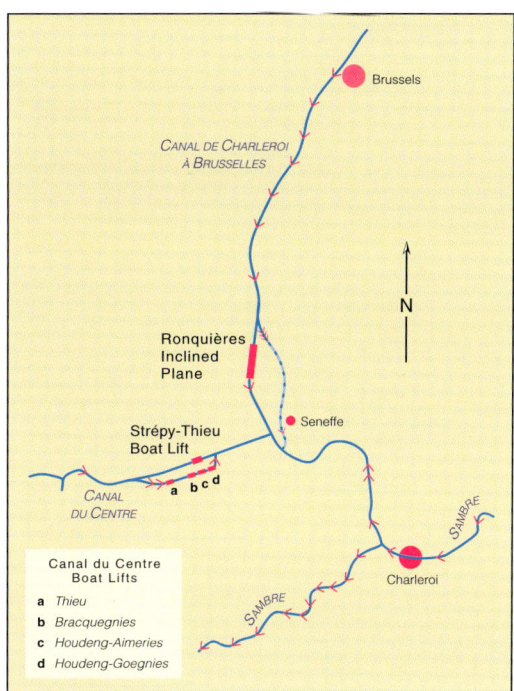

Fig. 169 The location of the Canal du Centre lifts and the Ronquières incline.

between the canal and caisson gates is filled with water by opening a sluice in the canal gate. Then both gates are raised slightly to allow the water level to stabilise before they are raised completely.

The design for the oldest of the four lifts, No. 1, Houdeng-Goegnies, followed Clark's patent and during construction there was an exchange of information with the French engineers who were building Les Fontinettes.

Fig. 170 The old and the new: hydraulic lift No. 4 at Thieu, with the counter-balanced Strepy-Thieu lift in the distance.

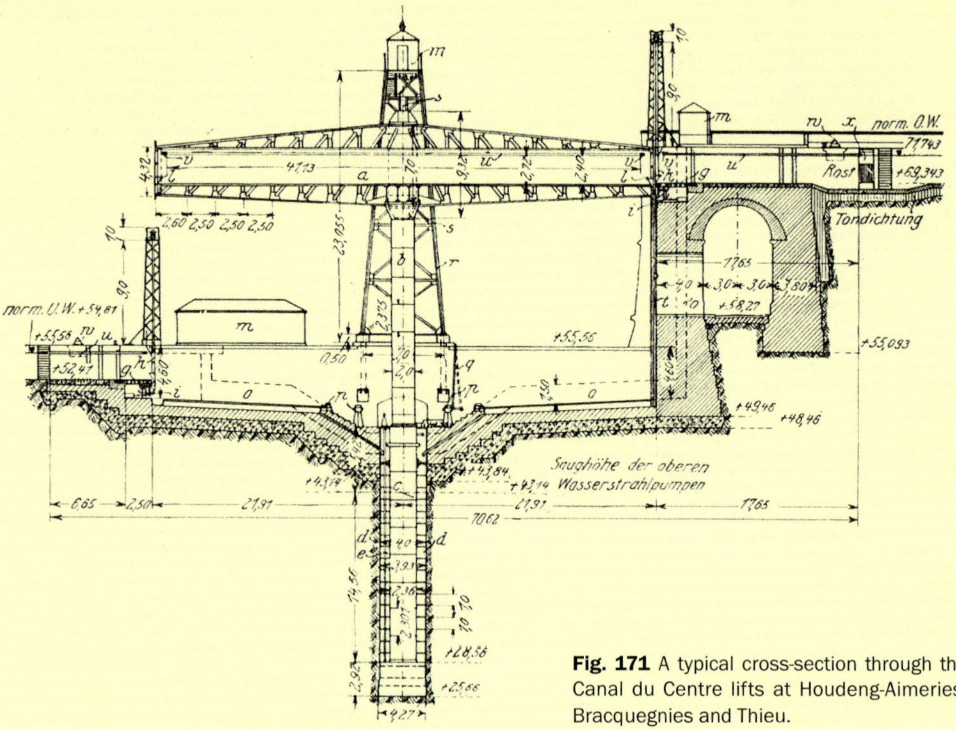

Fig. 171 A typical cross-section through the Canal du Centre lifts at Houdeng-Aimeries, Bracquegnies and Thieu.

Construction in Belgium was undertaken by the Cockerill Company from Seraing. Leopold II, King of the Belgians, ceremonially opened the Houdeng-Goegnies lift on 4th June, 1888.

The steelwork and mechanical engineering for the lifts at Houdeng-Aimeries (No. 2) and Bracquegnies (No. 3) were begun in 1909 and they were completed during the First World War. Through traffic on the Canal du Centre began on the 27th July 1917 with the opening of the Thieu lift (No. 4), which was completed by the then German occupying power.[131]

It took between 15 and 20 minutes to pass one of the lifts as, in contrast to Les Fontinettes, the Belgian lifts were provided with capstans. A single operation of a caisson lasts about $2^{1}/_{2}$ minutes and is smooth and noiseless. During the whole process, almost all you can hear is the turning of the gate guide rollers and the splash of the gates on entering the water.[132]

Three of the lifts are still in full operation but that at Houdeng-Goegnies (La Louvière) has recently been damaged, possibly as the result of a failure of a hydraulic seal. However, it is expected that they will all survive as a tourist attraction and working monument when the new canal and lift at nearby Strépy-Thieu (see page 129) opens. There is a well marked footpath along the canal to all four lifts and it is possible to arrange a trip on a boat, naturally including a passage through the lifts.

Fig. 172 The Pelton wheel drive for the hydraulic pump inside the power house at No. 3 lift, Bracquegnies.

Fig. 173 In January, 2002, there was a major accident to the Canal du Centre boatlifts, on the original lift at Houdeng-Goegnies. Due to a problem with the hydraulic system, the upper caisson began to descend, causing the lower caisson to rise just as a boat was leaving. The boat was severely damaged by the gate framework and by its stern being lifted from the water by the rising caisson. The weight of the boat held down one end of the caisson, and resulted in the distortion which can be seen in the photograph

Peterborough and Kirkfield lifts

The world's largest hydraulic lifts are on the Trent-Severn Waterway in Canada (see map on page 62), with that at Peterborough entering service in 1904 and that at Kirkfield in 1907. The lifts differ both in their rises, Peterborough 65 feet (19.80 metres) and Kirkfield 49 feet (14.9 metres), and also in their constructional materials and the general arrangement of their caisson guides.

At Peterborough, where the engineer was Richard B. Rogers, the caissons are designed for boats of 800 tons. They rest on 32 feet (9.75 metres) high latticework carriers and move vertically between three massive 100 feet (30 metres) high central guide towers and are also restrained by an 80 feet (24.4 metres) high and 40 feet (12.2 metres) thick upper canal wall. In contrast, the caissons at Kirkfield run between three steel towers which are linked by overhead bridging latticework.[133]

The caissons of both lifts have a usable length of 140 feet (42.67 metres), a width of 33 feet (10.06 metres) and a minimum water depth of 7 feet (2.13 metres). They need an additional $8\frac{1}{4}$

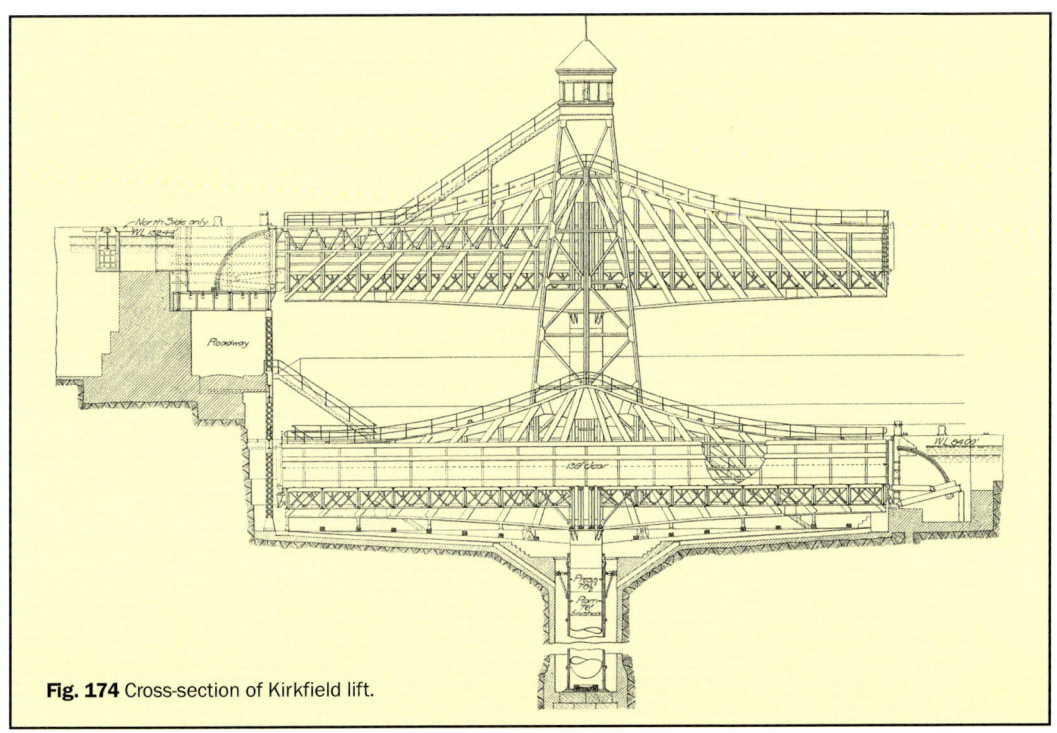

Fig. 174 Cross-section of Kirkfield lift.

Fig. 175 The Peterborough lift, Canada, in 1904, shortly after entering service.

inches (21 cm) of water in the upper caisson to make them move, corresponding to about 144 tons. The total weight of each caisson, about 1,700 tons, is transferred by a 9 feet (2.7 metres) high plate girder to the piston. This has an outer diameter of $7^1/_2$ feet (2.286 metres) and is made from cast iron sections $5^1/_4$ feet (1.6 metres) long with a wall thickness of $3^1/_4$ inches (83 mm). The cast steel cylinder, of inside diameter 7.7 feet (2.35 metres), operates at a pressure of about 600 lbs/in^2 (42 bar). However, it can withstand a pressure of up to 2,000 lbs/in^2 (141 bar). Both

Fig. 176 A recent view of Kirkfield lift.

the construction and operation of the gates deviates from those previously built. In order to maintain maximum navigable depth, folding gates are used both for the caissons and for the canal. They are hinged close to the bottom of the water. The gates had wrought-iron airtight tubes to give them buoyancy and they were powered by three-cylinder hydraulic engines supplied from a hydraulic accumulator. New, simpler welded aluminium gates were fitted in the 1960s, when the hydraulic engines were also replaced by oil-hydraulic cylinders.[134] In operation, the canal gate is folded down first and the caisson gate is then folded down on top, the caisson gate being slightly narrower than the canal gate. The gap between caisson and canal is sealed by inflatable rubber tubes.

Hydraulic lifts larger than these Canadian ones are unlikely to be built. It would require the caissons to be supported by two or more pistons which would cause problems with the distribution of power and keeping the caisson horizontal.

Anderton lift restoration

However, a 'new' hydraulic lift has entered service in 2002. Because of structural corrosion, Anderton lift (see pages 103 and 119) was taken out of service in 1983. Three years later, the Anderton Lift Development Group was formed, which worked with British Waterways, the waterway authority, to achieve the goal of returning this important industrial monument to working order. Because a large number of visitors are expected to come to the restored lift, some improvements have been made to the surroundings, such as the provision of picnic areas, landscaping, footpaths and car parks.[135]

There was much discussion about how the lift should be restored. Should it be retained in its twentieth century form as a counter-balance lift or as originally built as a hydraulic lift. In order to reduce the weight on the structure, it was decided to revert to the original hydraulic operation, but with new hydraulic pumps, cylinders and pistons. As with most modern hydraulic installations, oil is the working fluid, rather than water. On the lift, the framing added in 1908 is retained, together with the gearing, but the counter-balance weights have not been refitted and this has reduced the load on the structure by 540 tons. The lift was completely dismantled for restoration and it re-entered service in April 2002.

Counter-balanced Lifts

The principle of the counter-balanced lift, where a water-filled caisson is moved vertically and

Fig. 177 A view of Anderton lift after restoration.

counterbalanced either by a second caisson (twin lift) or by counterweights, probably dates back to 1794 when a description of a twin lift was published in Edinburgh by Dr. James Anderson. The two caissons were connected by chains which ran over large wheels and in order to balance the change in weight caused by the variation in the length of chain on either side as the caissons moved, secondary chains were fitted under them. The lift was to have been manually operated.[136]

Barrow Hill lift

The first lift similar to Anderson's scheme was tried by James Fussell on the Nettlebridge branch of the Dorset & Somerset Canal at Barrow Hill. It had twin caissons six feet (1.83 metres) wide, a rise of 21 feet (6.10 metres) and could be used by 10 ton boats. It was tested on the 6th September 1800 and gave full satisfaction, so five more lifts were ordered. Work began in 1801, but the canal company soon ran out of money and they were not completed. Traces of the lift chambers could still be found in the 1970s.[137]

Tardebigge lift

Somewhat more fortunate was the lift built for experimental purposes at Tardebigge top lock on the Worcester & Birmingham Canal. It was soon replaced by a lock, the top one of 30 in a distance of only 4 km, the greatest number in a single flight in Britain.

This lift had, in contrast to Barrow Hill, a single caisson 72 feet (21.95 metres) long, 8 feet (2.44 metres) broad and 3.5 feet (1.07 metres) deep, counter-balanced by bricks stacked on wooden frames. Balance was completed by hanging chains under the counterweights and the caisson. The lift had a rise of 12 feet (3.66 metres) and came into service on the 24th June 1808. After the tunnel at Tardebigge had been completed on the 26th February 1810, a boat carrying 20 tons passed through the lift taking just 2½ minutes. During testing between the 25th February and the 16th March 1811 the lift is reported to have passed 110 boats in 12 hours. Afterwards, the engineer John Rennie reported

Fig. 178 A contemporary illustration of the lift at Tardebigge.

and suggested that the lift was too fragile for normal conditions, so the canal company decided against further lifts and built locks instead.[138] The lift seems to have closed in 1815, either just before or after the canal was completed, so it is unlikely to have had much use.

The Grand Western Canal lifts

Better known than those at Barrow Hill and Tardebigge are the seven vertical lifts on the tub-boat of section the Grand Western Canal.

A waterway link uniting the Bristol and English Channels was the dream of many canal schemers. An Act was obtained for the Grand Western Canal in 1796 and work began in 1810 on the middle section of the waterway serving the quarries at Burlescombe. By 1814 an 11 mile long, lock-free canal had been opened between Tiverton and Lowdwells. The continuation to Exeter and the English Channel was never begun, but a link to the Bristol Channel was made in 1827 with the opening of the canal between Lowdwells and Taunton and then using the Bridgwater & Taunton Canal. Since only a comparatively small amount of lime and

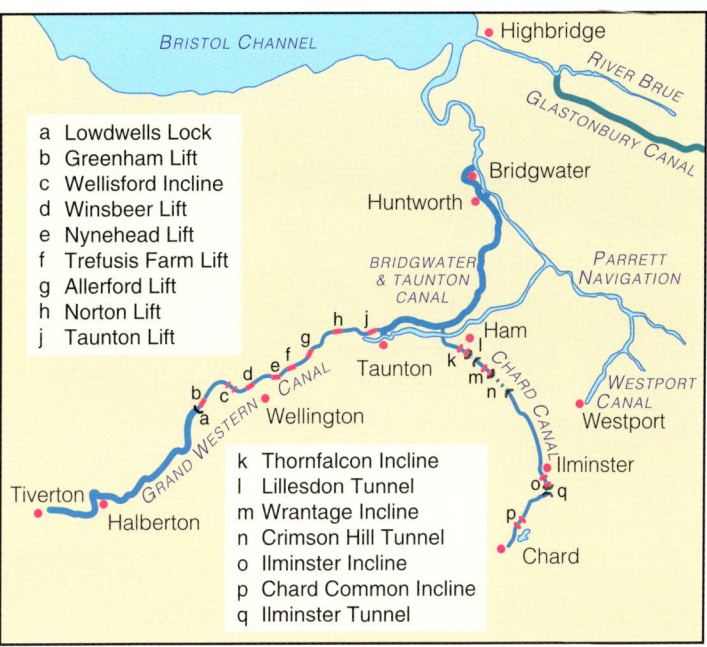

Fig. 179 Location of the Grand Western Canal and the Chard Canal.

a Lowdwells Lock
b Greenham Lift
c Wellisford Incline
d Winsbeer Lift
e Nynehead Lift
f Trefusis Farm Lift
g Allerford Lift
h Norton Lift
j Taunton Lift

k Thornfalcon Incline
l Lillesdon Tunnel
m Wrantage Incline
n Crimson Hill Tunnel
o Ilminster Incline
p Chard Common Incline
q Ilminster Tunnel

BRISTOL CHANNEL
Highbridge
RIVER BRUE
GLASTONBURY CANAL
Bridgwater
Huntworth
BRIDGWATER & TAUNTON CANAL
PARRETT NAVIGATION
Ham
Taunton
CHARD CANAL
WESTPORT CANAL
Westport
Wellington
GRAND WESTERN CANAL
Ilminster
Tiverton
Halberton
Chard

coal traffic was expected between Lowdwells and Taunton it was built as a tub-boat canal in contrast to the broad canal already completed. The tub-boats were 26 feet (8 metres) long and $6\frac{1}{2}$ feet (2 metres) wide and carried eight tons at a draught of $2\frac{1}{4}$ feet (70 cm) and a horse could haul a train of up to 20 such boats.[139]

Fig. 180 A recent view of the remains of Nynehead lift.

The canal would have needed thirty locks so to keep the quantity of feed water to a minimum and to do away with reservoirs James Green built seven vertical lifts as well the Wellisford incline mentioned earlier. The rise of the lifts varied between $12\frac{1}{2}$ feet (3.81 metres) at Norton Fitzwarren and 42 feet (12.80 metres) at Greenham and they each had two balanced caissons. They were built between 1830 and 1836.[140]

The upper and lower canals divided into two arms at the lift, each forming entrances to the caissons. The strong end walls were joined to two side walls and a central pillar. Openings in the central pillar and the lower end wall reduced the weight of the structure and helped to lighten the shafts in which the twin caissons moved up and down. In the central pillar were galleries and a staircase between the two levels.

The caissons were only slightly longer than the tub-boats and had a wooden floor and side walls. The ends were made from cast iron frames firmly bolted to the side walls and grooves in the frames allowed cast iron shutters to be fitted. A similar frame was fitted to all four canal entrances. The frames were covered by a mat made from tarred

117

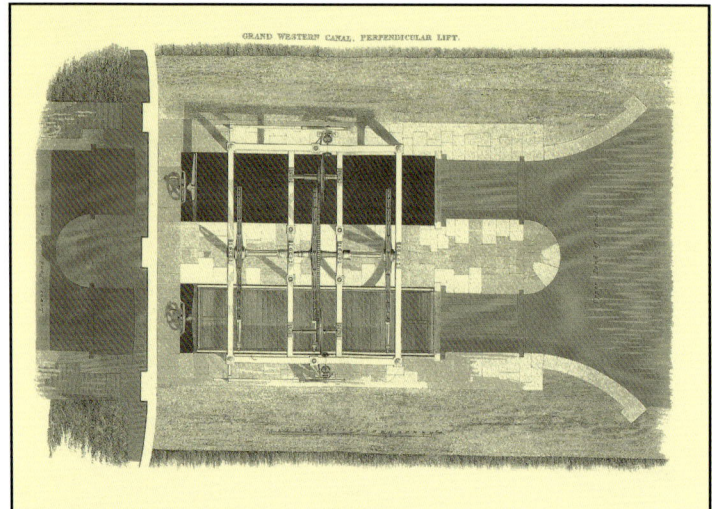

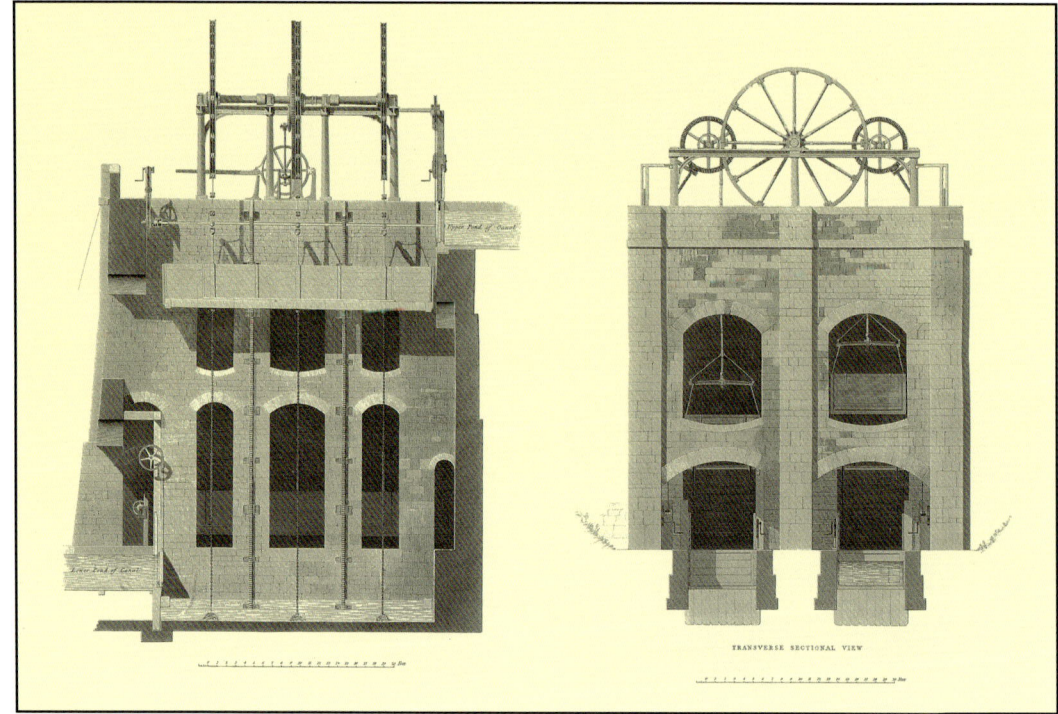

rope and were pressed against each other by wedges at the bottom and screw threads above to make a simple waterproof joint between the caisson and the canal. After the removal of the shutters, boats could sail in or out.

On top of the central wall there were three large, cast iron sprocket wheels of 16 feet (5.02 metres) diameter over which chains ran on which the caissons were hung by iron shackles. The grooves of the wheels in which the chains ran had flats which engaged with the chain links to prevent slippage. The middle wheel had a rim gear coupled on one side to two brakes. On the other side was a gear train turned by two cranks which allowed the caissons to be raised or lowered without adding any weight.

Variations in balance caused by variations in the length of the chains as the caissons rose or fell were prevented by three more chains hanging from the underside of each caisson.

Fig. 182 Another view of the remains of the Nynehead lift.

The bottom of the shafts had to be deep enough so that a caisson did not rest on these chains in its lowest position. Increasing the water level in the upper caisson by about two inches (barely 5 cm) gave an overweight of about one ton which sufficed to set the caissons in motion. They could then be safely regulated by the brakes.[141]

Green suggested that the time for passing through a lift was three minutes at Greenham where there was a rise of 42 feet (12.85 metres).[142] However other observers suggested elsewhere it was around $8\frac{1}{2}$ minutes. The difference may have been because the caissons at Greenham had pipes for drainage and so it was much easier to equalise the water level with the lower canal, though with comparisons it is always difficult to be exact. Despite overfilling from the upper canal and some leakage during lifting, the upper canal could receive up to 6 m³ of water. This was caused by the difference in displacement between loaded boats descending and empty ones ascending.[143]

The lifts carried little traffic, particularly after 1844 when the Bristol & Exeter Railway opened. The tub boat section of the canal with its lifts was closed in 1867; the section between Tiverton and Lowdwells followed in 1924. This reopened in 1971 when ownership passed from British Waterways to Devon County Council.[144] You can find out more about the Grand Western Canal in the Tiverton Museum where there is a model of a lift and plans and drawings in an archive. Of the lifts, some have disappeared and only a little masonry remains of the others. The best preserved is at Nynehead where the rise was 24 feet (7.32 metres). The land owner is conserving it and there is a Grand Western Canal Society who are working on other sites.[145]

Anderton lift reconstruction

Until the next new and much larger counter-balanced lift was built at Niederfinow, almost one hundred years were to pass. But the idea of this type of lift was not dead in the interim with the hydraulic lift at Anderton (see page 103) being reconstructed in 1908 as a counter-balanced lift. Up to May of that year, the lift had carried about 5 million tons of goods and

Fig. 183 Anderton lift around 1970.

Fig. 184 Work in progress on the reconstruction of Anderton lift in 1906-8.

annually there were about 15,500 boats using it and carrying some 200,000 tons of cargo. Reconstruction work started in 1906 and to keep traffic on the move only three short-term closures were necessary. This was both an operational and technological masterwork under the control of John Arthur Saner.[146]

The initial closure was from 6-00pm on Thursday 12th April to 6-00am on Monday 30th April 1906. Work was carried on day and night to install a $5\frac{1}{2}$ feet (1.70 metres) thick layer of concrete on the caisson chamber floor reinforced by I-section beams. In August, from 4-00pm Saturday 4th to 7-00am Monday 13th, the site was again drained. The frames for the new river gates were fitted and the concrete work completed to finish all the underwater works. Further work, such as the erection of the additional framework and the rope sheaves and gears on top of the lift, the fitting of the counterweights etc. could be completed without worrying over deadlines and traffic continued uninterrupted until Easter 1908. The last closure was from 16th April to 5th May, 1908, when all remaining works, such as removing the pistons and cylinders and fixing the counterweights to the caissons, were undertaken. Each caisson weighed 252 tons and was carried by 36 sets of wire ropes, each with a seven ton cast iron weight. On 5th May 1908 the new lift was tested for the first time and the reconstruction was completed at the end of July 1908.

The change in the system was to prove highly beneficial and it doubled the capacity of the lift. In the first year after reconstruction, 20,527 boats were passed by 17,285 operations of the lift.[147] The main goods transported were salt, coal, pottery clay and pottery as well as agricultural products.

Fig. 185 Anderton lift during dismantling in 2000. The gears in the foreground are now refitted to the top of the structure.

Canal transport declined rapidly in the 1950s and that using the lift ended in the middle of the 1960s. However, there was already an increase in pleasure traffic which wanted to use the lift. Unfortunately there were continual problems with corrosion of the structure and it was finally taken out of service in 1983. In 1987 the counterweights, the rope sheaves and the gearing were removed to reduce the weight on the lift, with all the disassembled parts being stored nearby. However, plans were soon being laid for restoration (see page 115).

Niederfinow lift

On 21st March 1934 this lift entered service on the Havel Oder Waterway, sometimes called the Großschiffahrtsweg Berlin-Stettin or the Hohenzollernkanal. Its dimensions exceeded all previous lifts.

The Havel Oder Waterway (HOW) had an important precursor over which traffic between the Havel and Oder had formerly passed. The Finow Canal was built between 1605 and 1620, but in the turmoils of the Thirty Year War it was

Fig. 187 The scale working model of the Niederfinow lift.

totally destroyed. Under Frederick the Great, between 1743 and 1746, it was rebuilt and by the end of the nineteenth century it was one of the most heavily used inland waterways. Boats up to Grossfinow size (41.5 by 5.1 by 1.6 metres, maximum load 250 tons) descended by 13 locks from the Havel to the old Oder at Liepe. When traffic on the canal reached almost three million tons per year, a further increase was impossible even though the locks had been doubled. Bypassing the old canal, the larger and more productive HOW was built between 1906 and 1914. This was designed for Plauer sized boats (65.0 by 8.0 by 1.75 metres, maximum load 650 tons) and descended 36 metres into the Oder valley at Niederfinow by a flight of four locks which replaced the smaller locks on the Finowkanal.[148]

On 1st April 1905 a Prussian waterway statute was passed authorising a second descent at Niederfinow. This time it was to be by lift. After numerous varied preliminary designs, some of which will be discussed later, finally the one which had been designed internally by the waterway administration was chosen because it contained the following essential features:

1   Vertical lift with the boat floating,
2   Dry lower section, ie caisson chamber dry,
3   The use of counterweights attached by wires to allow a perfect weight balance, including the weight of the wires as they moved,

Fig. 186 Research tower in Berlin-Dahlem used for testing full sized ropes and sheaves which were to be used on the Niederfinow lift.

UPPER HAVEL NAVIGATION
ODER NAVIGATION
Niederfinow Boat Lift
FINOW CANAL
ODER
HAVEL
HAVEL CANAL
HAVEL
BERLIN
HAVEL
SPREE
N
HAVEL
TELTOW CANAL
SPREE

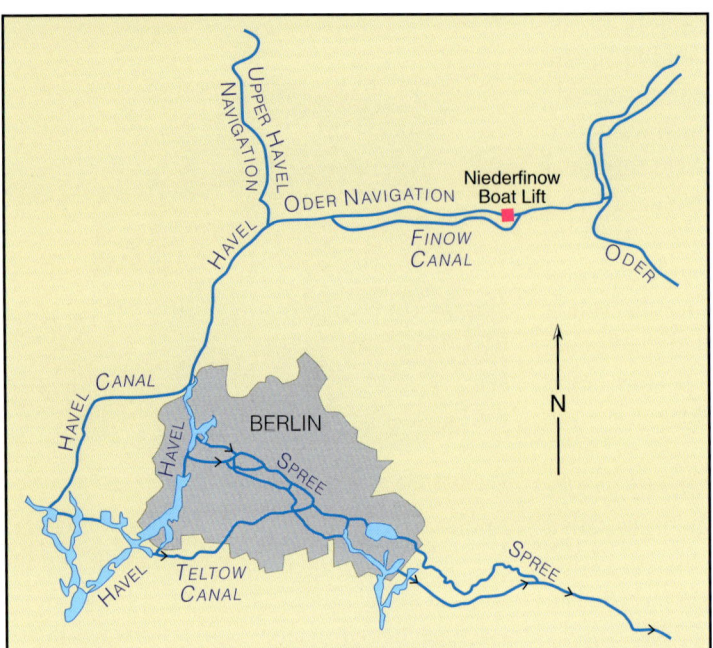

Fig. 188 Location of Berlin and the Niederfinow lift.

4 Vertical movement controlled by pinions working on four racks and mechanical linking of the gears by shafts which would guarantee that the caisson would remain horizontal,

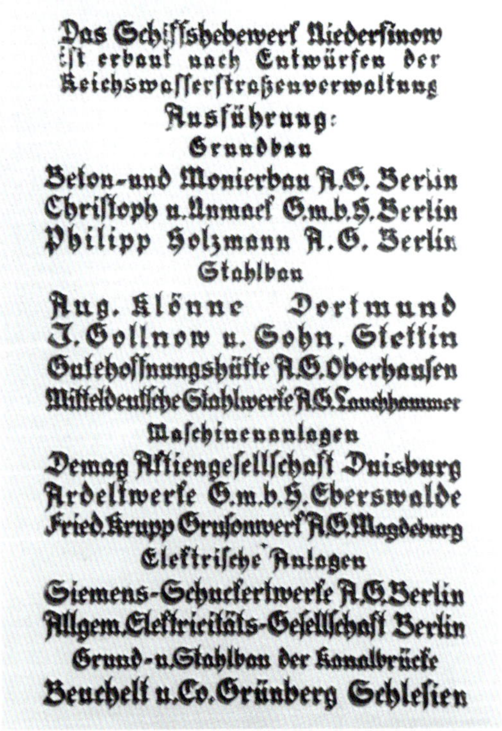

Das Schiffshebewerk Niederfinow
ist erbaut nach Entwürfen der
Reichswasserstraßenverwaltung
Ausführung:
Grundbau
Beton- und Monierbau A.G. Berlin
Christoph u. Unmack G.m.b.H. Berlin
Philipp Holzmann A.G. Berlin
Stahlbau
Aug. Klönne Dortmund
J. Gollnow u. Sohn, Stettin
Gutehoffnungshütte A.G. Oberhausen
Mitteldeutsche Stahlwerke A.G. Lauchhammer
Maschinenanlagen
Demag Aktiengesellschaft Duisburg
Ardeltwerke G.m.b.H. Eberswalde
Fried. Krupp Grusonwerk A.G. Magdeburg
Elektrische Anlagen
Siemens-Schuckertwerke A.G. Berlin
Allgem. Elektricitäts-Gesellschaft Berlin
Grund- u. Stahlbau der Kanalbrücke
Beuchelt u. Co. Grünberg Schlesien

Fig. 189 The commemorative plaque on Niederfinow lift.

5 A safety device providing security against high weight imbalances in either direction (caisson empty or overloaded) using four worm gears each turning between pairs of threaded bars over the complete height of the lift, as patented by Loebell.

The design was approved in February 1927 by the Akademie des Bauwesens (Academy of Architecture). Construction of the canal to the lower entrance to the lift had started by 1925.

Because of its much greater dimensions, the lift at Niederfinow could not be based on earlier designs and all structural elements had to be thoroughly tested. A research tower was built in Berlin-Dahlem (scale 1:1) to check the wires, rope sheaves and bearing housings, while drive, guides, safety mechanisms and gates were tested on a 1:5 scale model in Eberswalde. At about seven metres lift, this was larger than most of those on the Grand Western Canal.

The new lift was designed for 1000 ton boats, 80 metres long, 9 metres broad and 2 metres draught although the waterway was only designed for Plauer-sized boats of 650 tons. The caisson's measurements were 85 metres by 12 metres and 2.5 metres water depth. With normal water conditions, that is

Fig. 190 The lift in 1934.

normal water level in the upper canal and average water level in the lower, the lift has a rise of 36 metres. It is designed, however, to allow for an excess of up to 70 cm in the upper canal. Including an allowance for a wind surge of 40 cm, the maximum rise for the lift is 37.21 metres.

Built from St 52 steel, the empty caisson weighs 1,600 tons and 4,300 tons at normal water levels. It is restrained in both longitudinal and transverse directions by rollers and hangs on 256 wire ropes in groups of eight. Pairs of ropes are led over a double-grooved rope sheave of 3.50 metres diameter. In each group six of

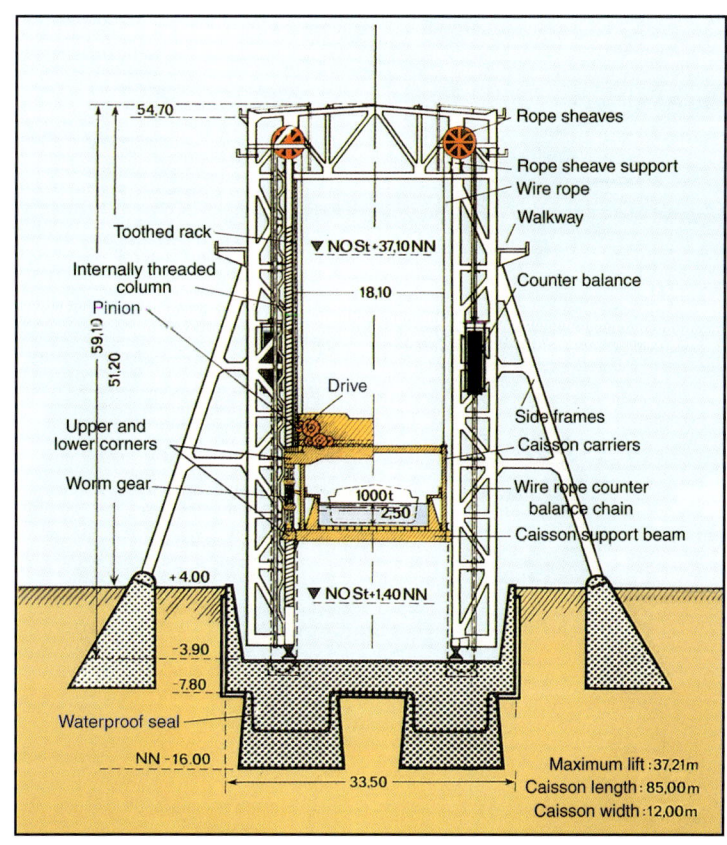

Fig. 191 A diagrammatic cross-section through the lift.

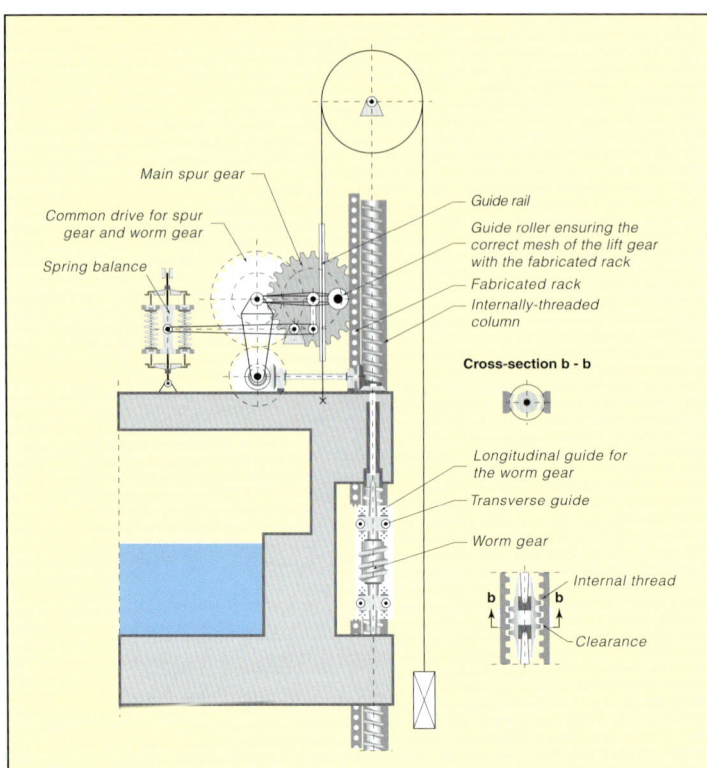

Fig. 192 Schematic cross-section through the main drive and safety mechanism.

Main spur gear

Common drive for spur gear and worm gear

Spring balance

Guide rail

Guide roller ensuring the correct mesh of the lift gear with the fabricated rack

Fabricated rack

Internally-threaded column

Cross-section b - b

Longitudinal guide for the worm gear

Transverse guide

Worm gear

Internal thread

b   b

Clearance

the eight wire ropes carry a counterweight running in a frame which guides and holds it if the ropes break. Each counterweight is made from six smaller weights of about 20 tons each and made from concrete and iron filings. The round braided wire rope has six strands with 37 wires of 2.4 mm diameter in a circle around a hemp core. Their weight is balanced by four flat-link chains of equal in weight which are fastened in a loop from below the caisson to below the counterweights.

Power comes from four Ward-Leonard DC motors, each of 75 HP (55.2 kilowatts), which drive four spur gears connected by shafts. The motors and their DC converter are situated in two machine rooms on the caisson.

Fig. 193 A view over the caisson showing one of the machinery rooms and the synchronisation shafts.

The average lift speed is about 12 cm/s, so that approximately five minutes are taken for each lift. The four motors also turn the safety worm gears; four 1.40 metres long screwed spindles with a core diameter of 0.78 metre, which are fitted to the caisson. They ensure that any large imbalance in load is transferred directly to the lift structure. The worm gears are fitted and turn between threaded bars — called Mutterbackensaüle (internally threaded column), which are built into the lift structure. In the normal working conditions the worm gears turn as the caisson rises or falls such that they do not touch the threaded columns but have a clearance of 30 mm. The main spur gears are held in mid-position by a sprung arm. When there is an imbalance of 30 tons on one of these gears, the prestressed springs give and it comes out of mesh. With a corresponding automatic increase in the spring tension, the clearance between the worm gear and threaded bars is reduced and if there is a sufficiently large deflection they touch. The worm gear engages with the threaded column and the caisson is automatically held firm, disengaging the other

Fig. 194 The drive to the rope sheaves.

gears which are still being driven by the motor. When a worm gear engages with the threaded column, it always results in the lift stopping or almost stopping.

The lift structure is about 60 metres high, 94 metres long and 27 metres broad and is made from St 37 steel. It is composed of three towers built alongside each other and held together by an overhead cross-braced double-jointed frame. This also carries the weight of the 128 rope sheaves.

In the western tower is the upper canal entrance and a pressurised water supply created by the lift's height. The central tower comprises

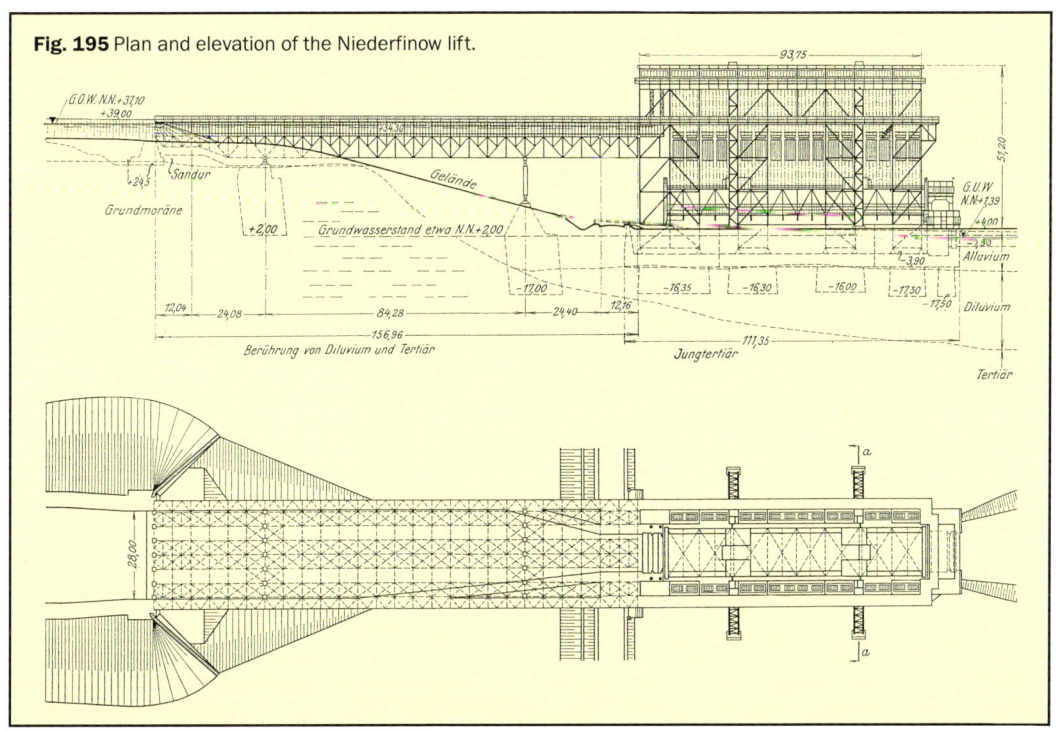

Fig. 195 Plan and elevation of the Niederfinow lift.

Fig. 196 One of the two rows of sheaves for the wire ropes which carry the caisson and counter-balance weights.

two parts and carries the fabricated toothed racks and the columns for the worm gears and threaded columns. Side movement is prevented by four buttresses. The caisson is fitted with lift gates at either end and at the top of the east tower is a curved buffer beam on a level with the gates which prevents the outer one from being damaged by a moving boat. Close to the upper canal gate is a safety gate, which in case of a breakdown or during repair work can isolate the lift. There is no safety gate on the lower canal and stop planks are used in case of repair or accident. When the caisson is in motion, all four gates are locked.

At the upper or the lower level the caisson is held against the entrance gate by means of a locking device comprising two levers fitted to the lift structure at either side of the caisson. These are pressed against shoulders on the caisson by feed rods driven by electric motors. After the gates have been locked together, the 10 cm wide gap between the caisson and the canal is sealed by pressing a U-shaped frame around the end of the canal by means of 14

Fig. 197 A recent aerial view of the Niederfinow lift. The old lock flight is in the valley to the right of the lift.

electrically-powered, sprung screw presses. Water fills the space between through pipes from the canal. At the upper level, it is emptied by a downspout into the lower canal. At the lower level it empties into a sump from where the water is pumped back into the lower canal.

Due to unfavourable ground conditions, it was not possible to build the lift into the slope of the Oder valley and the upper lift approach is across a 157 metre long, steel aqueduct supported on seven lattice girders.

On the lift it takes 20 minutes, including entering and leaving, to pass from one level to the other compared to over an hour on the old lock flight.[149]

For over 40 years, Niederfinow was the world's largest vertical lift. It still operates extremely reliably today because of its skilful design and construction and its careful maintenance. By its 60th anniversary in March 1994, it had carried some 127 million tons of goods, mainly building materials, coal, fertiliser, iron ores and scrap.[150] However the Niederfinow lift's days as the only descent to the Oder valley (the Niederfinow lock flight has been out of service for many years) are numbered. The improvement of the HOW to a class Va waterway for boats 110 by 11.4 metres, though with a reduced loaded draught of 2.20 metres, and the fact that the lift is carrying its maximum volume of goods (about 3 million tons annually) means that a new lift is required. It is likely to be a counter-balanced vertical lift with a caisson 115 metres long and 12.5 metres wide, though locks were also considered.[151]

The lift has always been an attractive place to visit. Visitors were even encouraged in the 1930s during construction and people were shown around on walkways above the construction site. Today, there is a visitors' platform where you can see the caisson in motion. You can also take a passenger boat for a trip through the lift. A walk around the old lock flight, now designated an industrial monument, and the 1746 entrance lock to the Finow Canal at Liepe a short walk away below the lift should not be forgotten.

Unfortunately, for the many visitors to the lift, there is no information centre. All those who want to discover more are recommended to visit the Inland Shipping Museum at Oderberg, some seven kilometres away.

Scharnebeck lift

On 5th December 1975, after about six years under construction, the Scharnebeck lift at Lüneburg entered service. The lift is part of the Elbe Lateral Canal which connects the Elbe at Artlenburg, above Hamburg, with the Mittelland Canal west of the lock at Sülfeld and forms a link for boats of 1,350 tons capacity between the port of Hamburg and the western Germany unaffected by variations in water level on the central Elbe (see map on page 100). The canal is about 115 km long and has a rise of 61 metres of which 23 metres are at Uelzen lock and 38 metres at the lift. It is about two metres higher than Niederfinow.

It is a double lift, its two caissons having a length of 100 metres, width 12 metres and water depth 3.50 metres. The weight of a caisson filled

Fig. 198 One of the caissons at Scharnebeck.

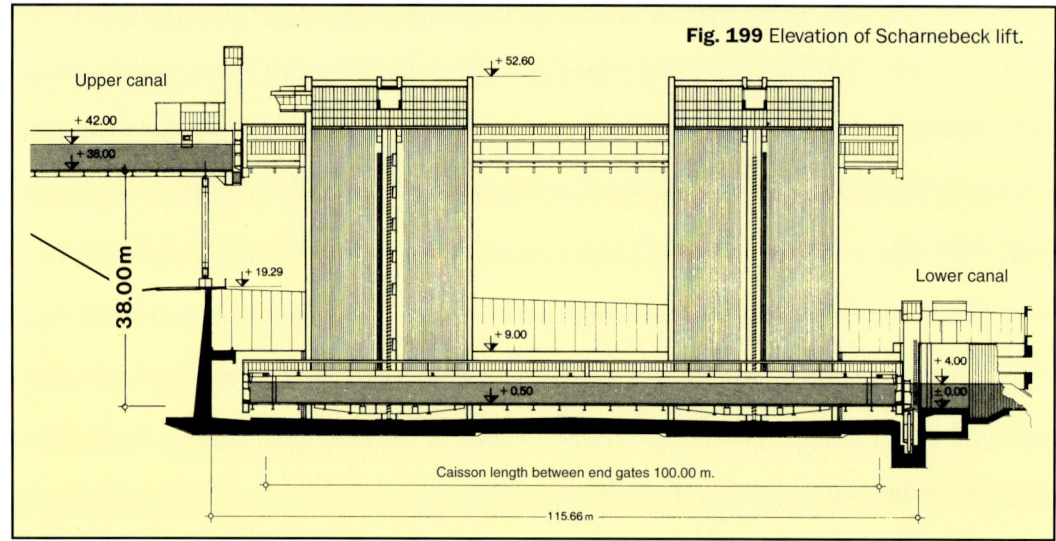

Fig. 199 Elevation of Scharnebeck lift.

Upper canal
+ 42.00
+ 38,00
38.00m
+ 52.60
+ 19.29
+ 9.00
+ 0.50
Lower canal
+ 4.00
± 0.00

Caisson length between end gates 100.00 m.

115.66 m

with water is about 5,700 tons and this is counterbalanced by eight weights each made from eight steel slab weights and 224 high-density concrete discs each weighing about 26.5 tons. There are 28 discs to each weight. The lift consists of four concrete guide and counter-weight towers for each caisson. There are 240 eight-stranded long-laid wire ropes, diameter 54 mm, for each caisson and these are led over double-grooved rope sheaves of diameter 3.40 metres and connect a caisson to its counter-weights. The total load of about 11,400 tons is carried by eight counterweight chambers in which a steel beam framework sits on the 0.40

metre thick reinforced concrete inner wall of the guide towers. In the towers are the toothed drive racks and the threaded shafts of the safety mechanism as well as staircases and elevators.

Each caisson rests on two support frames on which the four 160 kilowatt AC drive motors are also housed. They drive pinions engaged with the vertical toothed racks through gear trains. The four motors are connected by a drive-shaft system. With a speed averaging 14 m/min, it takes approximately three minutes for the 38 metre change in level.

Synchronized with the four drive pinions are threaded nuts which revolve around spindles

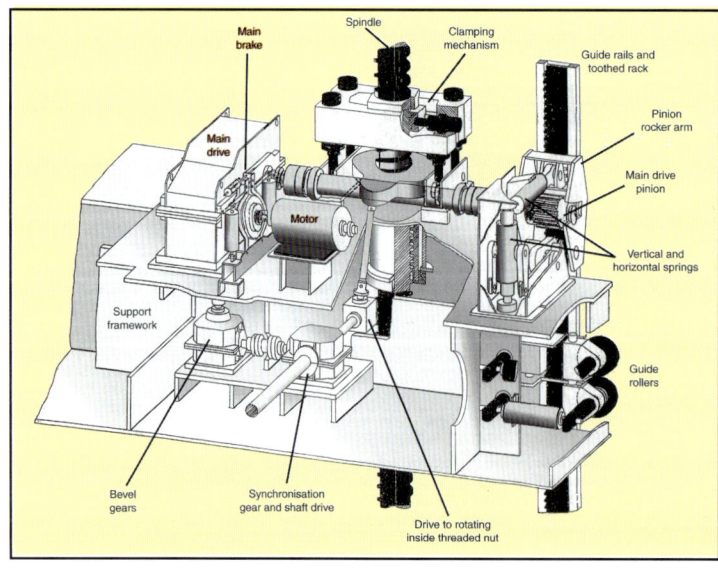

Fig. 200 Caisson drive mechanism on the Scharnebeck lift.

Fig. 201 A view of Scharnebeck lift in 1995.

fixed in the towers. There is 30 mm clearance between spindle and nut. If the drive pinions become overloaded due to an accident, the drive motors are automatically switched off and the nut presses against the upper or lower flank of the threaded spindle according to direction of the accidental load. The caisson can hang safely on the four spindles.[152] A letter from the WSA (the waterway authority) at Uelzen states that the spindles serve '...*no longer exclusively for safety purposes, but are also used operationally. A recently designed caisson stopping device avoids the problem that the additional vertical loads on the spindles due to levelling of the caisson and opening of the caisson and canal gates were not originally considered. With the new device, over-loads of the drive system are avoided.*'[153]

The upper and lower canals and the caissons are closed by lift gates. There is no drive to the caisson gates which are coupled to the canal gates by a latch and the two raised together.

The connection between the caisson and canal is made by a telescopic gap-sealing frame.

After the sealing frame is locked and the gap between both gates filled, the gates can be opened. Any variation in water level on the lower canal, which can be up to 4 metres, is allowed for by a movable frame in which the lower gate is suspended.

All operations are controlled from a central control room. The boats are directed to enter by signals and, after entering, all operations including the opening or closing of gates are fully automatic.

The lift is a magnet for visitors and there is a large information pavilion with displays on the theme of waterways and their development. Unfortunately access to the lift itself is not as good as at Niederfinow.

Strépy-Thieu lift

The construction of a counter-balanced lift on the Belgian Canal du Centre was begun in the 1980s. Enlargement of the canal for 1,350 ton boats began in 1971 and in 1982 construction started on a new canal and lift at Strépy-Thieu parallel to that with the old hydraulic lifts. The new 73.15 metre double lift will replace the four

129

Fig. 202 Strépy-Thieu in 1990 during the construction of the machine room and caisson guides.

old lifts and two locks. The entire construction, 130 metres long, 81 metres wide and a total rise of 117 metres, is on a tremendous scale.

The main structural works can be split into following components:

1 The concrete foundation slab. This is 60 metres under the original ground level on a thick stratum of sand and transfers the total weight of about 200,000 tons to the ground,

2 The central tower of reinforced concrete in the middle of the slab and a steel frame on either side of it. These transfer the loads from the caissons and counterweights to the foundation slab. In the central tower are lifts, storerooms and staircases,

3 The 20 metre high machine house and the main control room and viewing gallery for visitors.

Each caisson is 112 metres long, 12 metres wide and has a water depth of 3.35 to 4.15 metres according to the water level in the canal. Their weight varies between 7,200 and 8,400 tons. Eight 980 ton counterweights balance the weight of the caisson and are coupled to it by 144 steel cables of diameter 85 mm which run over drums in the machine room.

Each caisson is powered by four 500 kilowatt electric motors which are connected to the drive shafts by gears. The shafts are located over the side columns and at the edge of the central tower and are coupled for synchronisation. The shafts turn the cable drums supporting the caisson and balance weights through a second

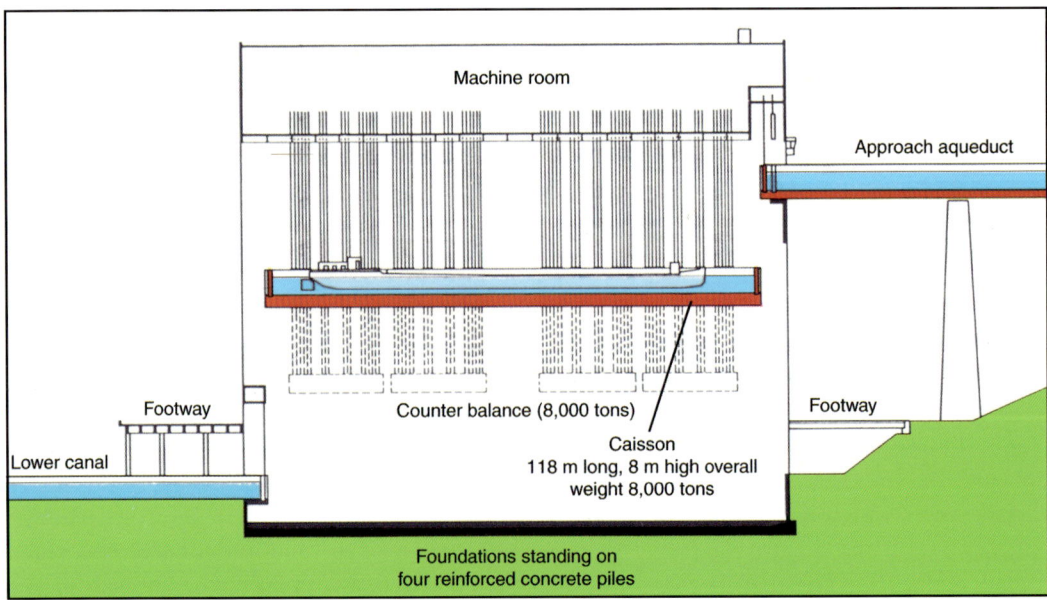

Fig. 203 Schematic longitudinal section of the Strépy-Thieu lift. The caisson is long enough for either a standard 1,350 ton boat or a 2,000 ton push-tow unit.

Figs. 204 & 205 The caisson being tested and the lift's machine room, both in 2000.

set of gears.

The caissons and canal sections are closed off by lift gates, which are raised or lowered together. On the upper canal, two lift gates are provided for safety. The caissons are connected to the upper canal by two 168 metre long steel aqueducts constructed in four sections.

One operation of the lift takes about 38 minutes. Some 28 minutes are taken up by entering and leaving the caisson, four minutes for opening and closing of the gates and six minutes for the raising or lowering the caisson. The average speed is about 20 cm/s.[154]

The lift was fully tested on the 6th November 2001 and will enter service as soon as the motorway aqueduct on the upper canal is completed. Visitors can find out more at the information centre where there are displays about the lift's design and construction as well

as about the nearby hydraulic lifts.

China

For the navigation section of the Three Gorges Dam project in China, a lift with a rise of 113 metres is being built, the caisson of which will be 120 by 18 by 3.5 metres.[155]

In fact, there are several operational counter-balance lifts for allowing boats to pass dams in

Fig. 206 The lift during construction in May 1997.

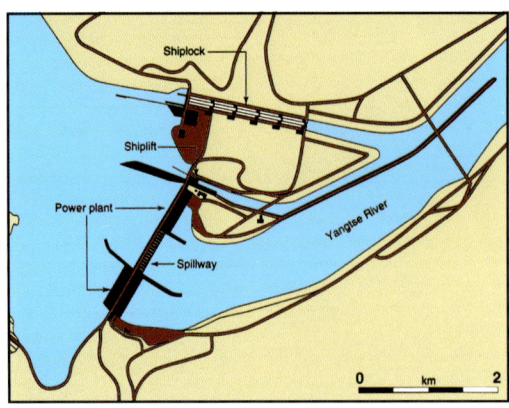

Fig. 207 The site of the Three Gorges project showing the hydro-electric plant, the lift and the lock flight.

China and more are proposed. At Shuikou, on the Mingjiang River, a 59 metre lift entered service in 1998. The caisson is 112 metres by 12 metres and is designed for two 500 ton boats. Also in operation since 2000 is the Yantan lift on the Hongshui River. This has a lift of 68.5 metres, with a caisson 40 metres by 10.8 metres for 250 ton boats.

Two more lifts are being built on the Geyehan Dam system, one of 42 metres lift and the other of 82 metres. They are designed for 300 ton boats, with caissons 42 metres by 10.2 metres. However, it is reported that there have been problems with these lifts which have put back their development. Another 300 ton counter-balanced lift is proposed at Gaobazhou on the Qingjiang River with a lift of 40.3 metres.

The largest of these new lifts will be on the controversial Three Gorges project previously mentioned. As well as the lift, there will be a flight of five locks, each 280 metres long and 35 metres wide, with a water depth of 5 metres, designed to handle 10,000 ton barges. The lift will be a single-stage vertical lift capable of carrying a 3,000-ton passenger or cargo vessel. As a result of the new dam, shipping is expected to increase on the central Yangtze from around 10 million to 50 million tons annually, with transportation costs cut by 30-37 percent.[156]

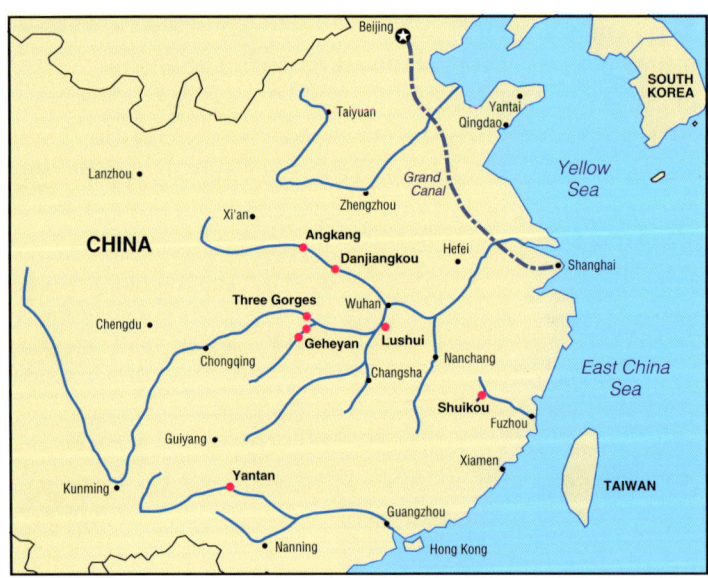

Fig. 208 Location of the various lifts in China.

Unusual types of lift

During planning for the Niederfinow lift, there were many suggestions and proposals which seem unusual or even fantastic today. Some of these ideas, fundamentally different from the present types of lift, are considered below.[157]

Cylindrical Lifts

In 1898, during preparations for the design of the enlarged waterway route from Berlin to Stettin, three firms had drafted schemes for transverse inclined planes at Niederfinow. After the Prussian waterway statute of 1st April 1905 was passed, a competition was held in 1906 which resulted a proposal by MAN for a lift in the form of a rotating cylinder or drum. This idea first appeared in a competition a few years earlier for a 36 metre lift at Prerau on the proposed Oder Danube Canal. The Austrian engineer Umlauf was awarded second prize for this scheme. The design by MAN proposed a revolving cylinder lift with a length of 68 metres and a span of 51 metres. Its lowest sector would have been in the water of the lower canal and it would have had a structural weight of 5,400 tons and caissons weighing 10,600 tons. These were 12 metres in diameter, closed at either end by gates and fixed. The upper canal also had to be closed off, though a gate was not necessary for the lower canal. The drum could be rotated through 180 degrees in either direction.

'The drive shaft, at the same level as the lift cylinder axle, is supported on either side by arms which allow a small movement to and fro. The axle for the arms has a pair of main drive gears (forward and reverse). These engage with the secondary gears which are also engaged with the toothed spur gear segments around the front of the cylinder. During rotation of the drum, it is controlled by these pinions.'(Ellerbeck) [158]

The Academy of Architects considered the design's main outline as *'a simple and very efficient lift. However the manufacture and maintenance of the exact circular form of the cylinder under the influence of its weight, the rotation and uneven thermal expansion would be difficult because of its large size, so there is no*

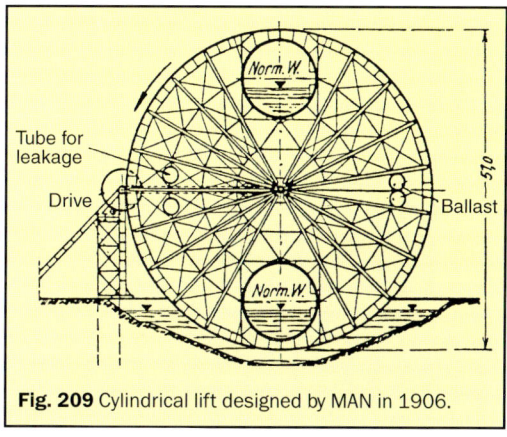

Fig. 209 Cylindrical lift designed by MAN in 1906.

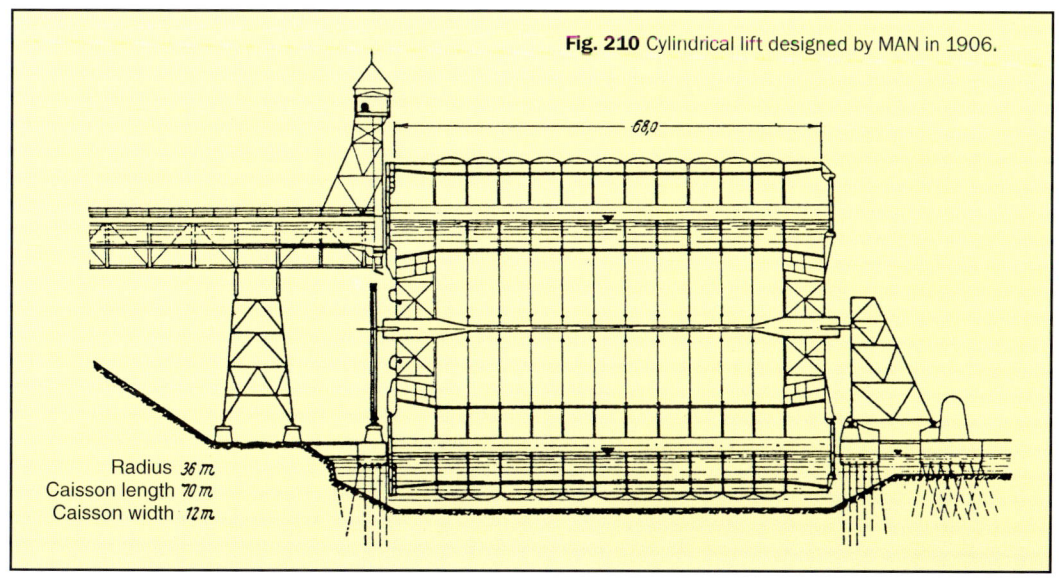

Fig. 210 Cylindrical lift designed by MAN in 1906.

Radius *36 m.*
Caisson length *70 m.*
Caisson width *12 m.*

Fig. 211 The Falkirk Wheel in April 2002.

assurance of the full engagement of the pinions and jamming might occur. One possibility would be to redesign the drive machinery and to roof over the whole structure as protection against wind and weather (which would be quite costly). Further thought needs to be given to connecting the caissons and the upper canal [a new connection was proposed by MAN for coping with large variations in water level] *Finally; more safety must be provided for the boats in the caissons when they are turning.'* MAN eventually gave up this proposal for a cylindrical lift.[159]

However, a lift of this type has been built in Scotland, where it links the Union Canal to the Forth & Clyde Canal and replaces part of a lock flight which had been destroyed earlier in the twentieth century.

The design for the Falkirk Wheel, the world's first operational rotating boat lift, was unveiled in December 1999. It is 115 feet (35 metres) high, 115 feet (35 metres) wide and 100 feet (30 metres) long and can carry eight pleasure boats at a time, depending on their length. Each caisson weighs 300 tons and is lifted 25 metres. A single trip will take about 15 minutes.

Construction involved 7,000 cubic metres of concrete, 1,000 tons of reinforcing steel, 1,200 tons of prefabricated steel and 35,000 square metres of canal lining. The Wheel is designed to last for at least the next 120 years and the cost of all the works is approximately £17

Fig. 212 A second view of the Falkirk Wheel in April 2002 during the final stages of construction and shortly before entering service..

million, including the lift, aqueduct, new canal link, basin and associated works.

Day visitors can experience The Wheel from special trip boats. On land, a new visitor centre provides a sensational vantage point from which to view the Wheel in action.[160]

Balance Beam Lifts

The following designs for balance beam lifts were suggested to give weight equalisation by lever action.

As early as 1906, the firm of Haniel & Lueg had produced a design for a lift using oscillating arms fitted to a floating cylinder. Eight 122 metre long arms formed the link between a floating cylinder, diameter 30 metres, and a caisson 68 metres long and 11 metres broad. The floating cylinder had two side reservoirs so that by pumping water from one to the other, the out-of-balance effect would make the cylinder and arms turn. Since the caisson was joined rigidly to the arms, it was always slightly out of level except in the mid-position, so that during the lift process the water was always subjected to a small sideways surge. As well as having some small safety risks, this design was much too expensive in comparison to others. Its great advantage, as opposed to the balance beam lifts to be described next, was that it did not need a main axle bearing. These had to withstand the weight of the whole moveable part of the lift, several thousands of tons, and were always the weak point in the design of both cylindrical and balance beam lifts, particularly when these were for large boats.

A design by MAN from the 1920s took the principle of the floating cylinder lift further by suggesting an upper and lower basin on the same axis into which the caisson would be immersed. The floating drum supporting the caisson would be moved sideways after the caisson had been lifted out of the water. It could then be rotated until it was at the correct height for the caisson to be repositioned over the basin into which it was to be lowered. In effect, the caisson was lifted over a dry summit level, so no gates were necessary on the canal.

Lifts with arms fitted to floating cylinders were an interesting idea, but construction was never considered in earnest. The opposite was the case with balance-beam lifts with fixed bearings. Beuchelt & Co. of Grünberg, in which Beuchelt, Schnapp and Bruno Schulz were involved, patented a design for this type of lift in 1906. A similar one was envisaged for the development of waterways in Russia prior to the First World War when the Tzar was trying to modernise the country. Extensive canal development was proposed, including a Trans-Siberian Canal and a canal between the rivers Oka and Dniepr. The German design was further developed by Beuchelt & Co., and by 1912 had reached the stage where it could be constructed.

Their design was described as follows: *'Four plate-girder balance beams, 60 metres in length, are held together by two frameworks and are balanced on a massive substructure such that they can turn. From them are hung two caissons*

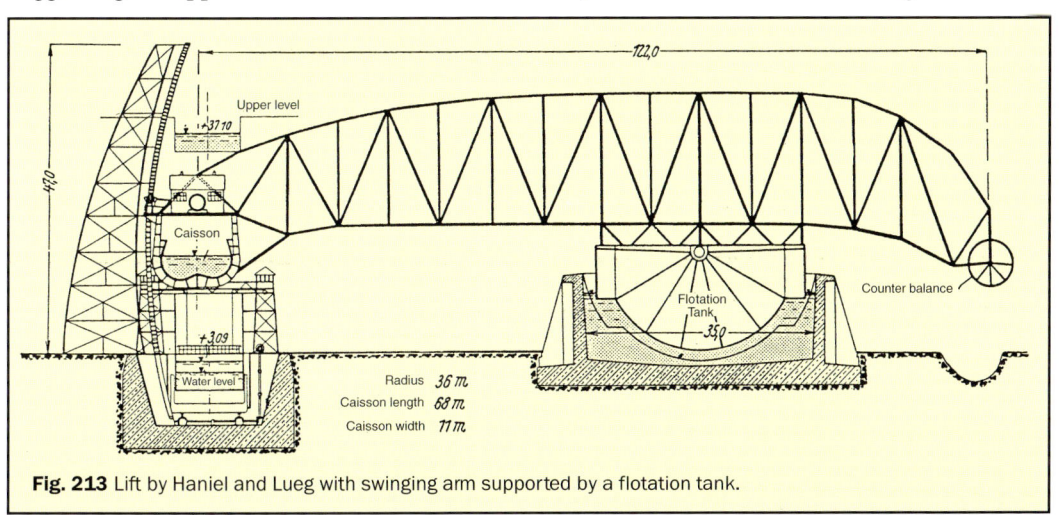

Fig. 213 Lift by Haniel and Lueg with swinging arm supported by a flotation tank.

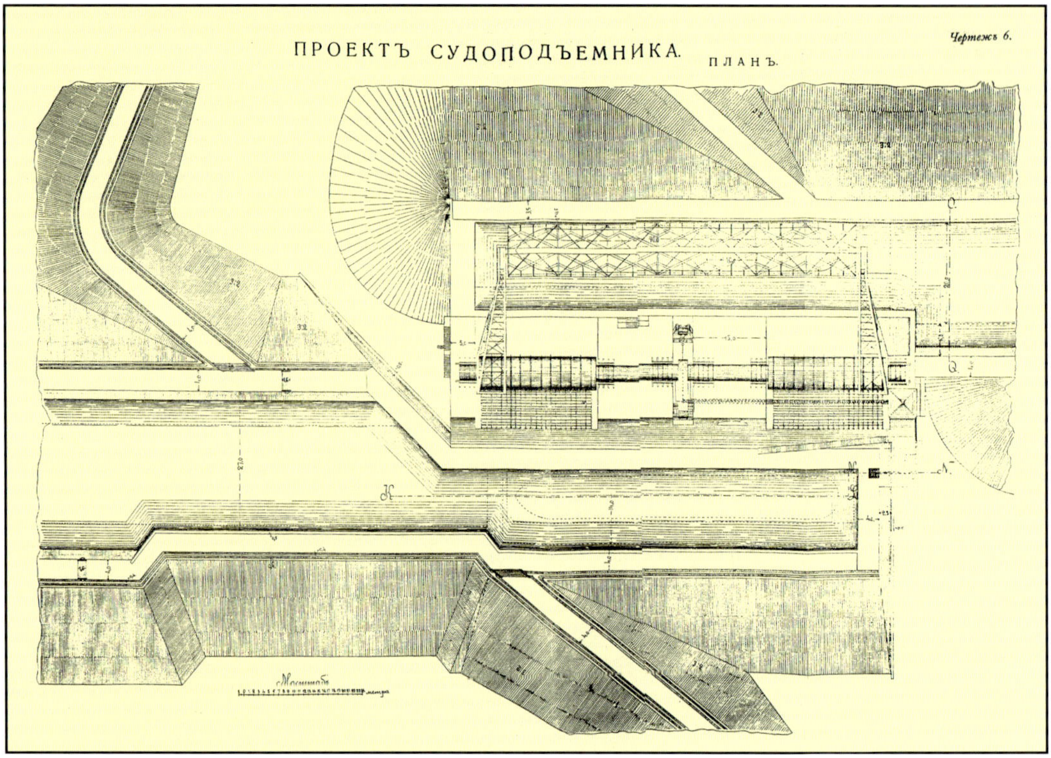

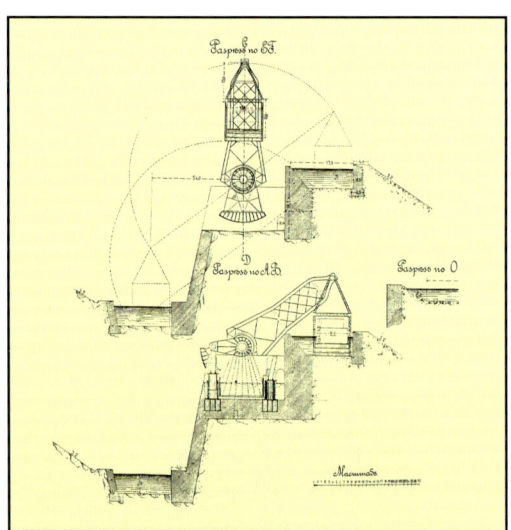

ПРОЕКТЪ СУДОПОДЪЕМНИКА. ПЛАНЪ.

Чертежъ 6.

Fig. 214 Design for a Russian balance beam lift.

connected by a linkage, although only one was used and the other replaced by ballast. The beams rested on four steel gudgeons of 70 cm diameter. A later suggestion was made by Privy Councillor Zimmermann in which they were made tubular and cross-wires inserted so that the exact position of the gudgeons could be found at any stage. This is of great importance as the bearings were to be adjustable. To regulate the movement of the caisson two damper paddles were proposed which were, in contrast to the 1906 design, connected by a rod and guided by a parallelogram linkage. In order to keep the caisson vertical, two guide bars connecting a second parallelogram linkage were used; pinions were used for the drive.'[161]

It was a competition design given the highest marks by the Academy of Architecture in 1912, mainly because of the safe control of the loads and the skilful braking by damper paddles of any imbalance. It was also praised by the waterway administration for the precise design. However, preparation was interrupted by the outbreak of the First World War and the initial works had to be stopped.

Technical difficulties to be overcome were in particular:

1 The high strain on the bearings, the size of which would have had to be increased later when the size of the waterway was raised to 1,000 ton standard. This would have required an increased size of caisson (each of the four

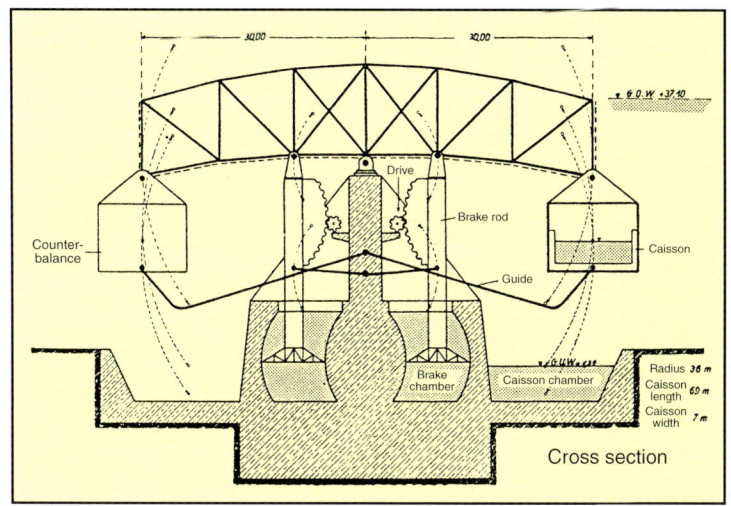

Fig. 215 Balance beam lift by Beuchelt with water damping, from 1906.

steel gudgeons would then have had to take a load of 2,600 tons),

2 The large mass of the swinging balance beams,

3 The high wind pressure on the structure,

4 The reliance for safety on exact alignment of the bearings which would have been complicated where there was an uneven ground.

These difficulties, but above all the search for a more economic solution, led ultimately to Beuchelt's suggestion of 1922. This proposed holding the caisson on both sides with balance beams and for it to rest on rolling axles (roller bearings). But this idea also gave cause for the reflection that guiding the caisson, especially at large load imbalances, was not ensured by pinions engaged with toothed racks. With this last variant, the design of balance-beam lifts ended and the waterway administration went ahead with the design and construction of the Niederfinow counter-balanced lift.

Looking at the success of the vertical lifts built over the last decades, both counterweight and flotation, it seems surprising that other designs have been suggested. Lifts, despite all their complexity and costliness, still provide admirable examples of ways of overcoming changes in level on inland waterways.

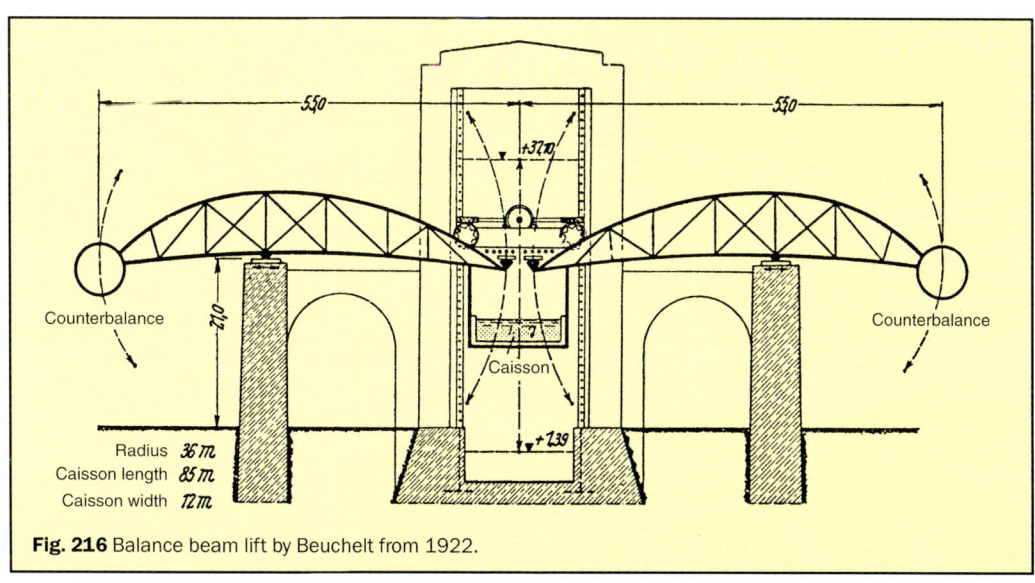

Fig. 216 Balance beam lift by Beuchelt from 1922.

Standard Reference Works

Carden, David, *The Anderton Boat Lift*, 2000, ISBN 0-9533028-6-5.

Dehnert, H., *Schleusen und Schiffshebewerke*, Berlin, Göttingen, Heidelberg, 1954.

Engels, H., *Handbuch des Wasserbaus*, Vol. 2, Leipzig, 1923.

Hadfield, Charles, *World Canals*, 1986, ISBN 0-7153-8555-0

Hagen, D., *Handbuch der Wasserbaukunst, Part 2, Die Ströme*, Vols. 3 & 4, 1857 & 1874.

Paget-Tomlinson, Edward, *The Illustrated History of Canal & River Navigations*, 1993, ISBN 1-85075-276-1.

Partenscky, H., *Binnenverkehrswasserbau - Schiffshebewerke*, Berlin, 1984

Schinkel, Eckhard, *Die Schiffs-Hebewerke der Welt*, 2001, ISBN 3-88474-834-3

Simons, H., 'Über die Gestaltung von Schiffshebewerken', *Mitt. der Hannoverschen Versuchsanstalt für Grundbau und Wasserbau*, Franzius-Institut der TH Hannover, 1957.

Tew, David, *Canal Lifts and Inclines*, 1984, ISBN 0-86299-031-9

Some Internet sites

Anderton
http://www.andertonboatlift.co.uk

Canada:
http://collections.ic.gc.ca/waterway/x.htm

Elblanski Canal:
http://www.ga.com.pl/elblag1.htm (also 2 & 3)
http://www.key.net.pl/firmy/kanale/index.html

Falkirk Wheel
http://www.falkirk-wheel.com

Foxton
http://www.foxcanal.fsnet.co.uk

General:
http://home.t-online.de/home/Guenter.Ringe/schiheb1.htm

Henrichenburg
http://www.route-industriekultur.de/prmaer/info/a05/a05.htm
http://www.industriedenkmal.de/henrich/henrich_text.html

Morris Canal
http://www.canalsocietynj.org/mcdata.html

The perpendicular lift performing an alternate task.

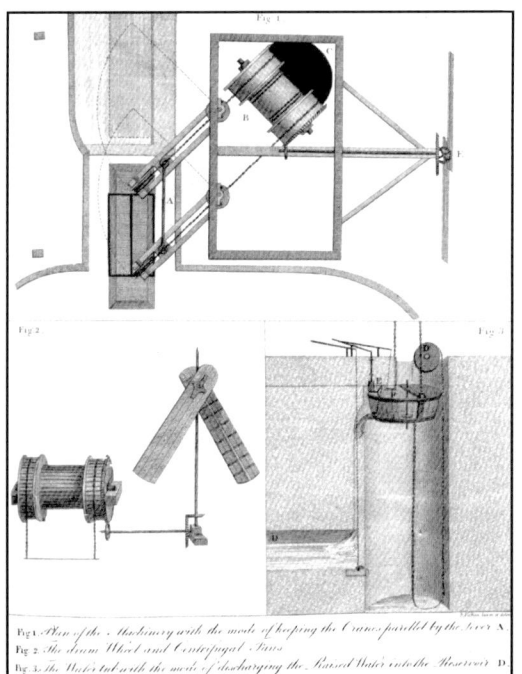

Fig 1. Plan of the Machinery with the mode of keeping the Cranes parallel by the lever A
Fig 2. The Steam Wheel and Centrifugal Fans
Fig 3. The Water tub with the mode of discharging the Raised Water into the Reservoir D

Location of the main European Lifts and Inclines.

1 Elbing Oberland Canal
2 Klodnice Canal
3 Orlik Dam Incline
4 Halsbrücke Boat Lift
5 Niederfinow Boat Lift
6 Rothensee Boat Lift
7 Scharnebeck Boat Lift
8 Henrichenburg Boat Lifts
9 Dutch Overtoom and Broekerhaven Boat Lift
10 Canal du Centre Boat Lifts & Ronquierres Incline
11 Les Fontinettes Boat Lift
12 Meaux Incline
13 Arzviller Incline
14 Montech Water Slope
15 Fonserannes Water Slope
16 Diolkos of Corinth

Left: Two illustrations from Fulton's book which is mentioned
on page 37.

1   Falkirk Wheel, Union Canal
2   Blackhill Incline, Monkland Canal
3   Worsley Incline, Bridgewater Canal
4   Inclines, Shropshire, Shrewsbury and Donnington Wood Canals
5   Anderton Boat Lift, Weaver Navigation
6   Foxton Incline, Grand Junction Canal
7   Tardebigge Boat Lift, Worcester & Birmingham Canal
8   Inclines, Kidwelly & Llanelly Canal
9   Combe Hay Boat Lift, Somersetshire Coal Canal
10 Inclines, Bude Canal
11 Inclines, Rolle Canal
12 Incline and Lifts, Grand Western Canal
13 Inclines, Chard Canal
14 Morwellham Incline, Tavistock Canal
15 Inclines, Ducart's Canal

Location of the main Lifts and Inclines in the British Isles.

Location of the main Lifts and Inclines in North America.

Legend on map:
1 South Hadley Canal
2 Morris Canal
3 Allegheny Portage Railway
4 Georgetown Incline
5 Shubenacadie Canal
6 Big Chute Marine Railway
7 Kirkfield Boat Lift
8 Peterborough Boat Lift

Illustration sources

H-J. Uhlemann collection: 5, 10, 13, 16, 23, 26, 27, 37, 41, 43, 47, 50, 51, 59, 60, 74, 82, 86, 89, 97, 98, 100, 101, 107, 110, 113, 114, 117, 118, 121, 140, 142, 149, 157, 162, 163, 164, 166, 168, 169, 170, 175, 179, 188, 189, 201, 206, 212, 213.

Ron Oakley: 58, 66, 108, 115, 116, 138, 145, 202.

Mike Clarke collection: 1, 2, 3, 17, 18, 19, 31, 32, 35, 38, 48, 51, 52, 54, 64, 69, 70, 80, 84, 85, 88, 92, 96, 102, 103, 104, 105, 106, 119, 120, 123, 126, 127, 128, 134, 150, 151, 153, 155, 158, 167, 172, 177, 180, 184, 185, 193, 194, 198, 204, 205, 207, 208, 214.

Hagen, G., *Handbuch der Wasserbaukunst*, 1874: 28, 52, 53, 62, 63, 65, 71, 72.

4   Eckoldt, M., *Schiffahrt auf kleinen Flüssen Mitteleuropas in Römerzeit und Mittelalter*, DSM 14, Oldenburg, 1980.

6   Uhlemann, H-J., *Berlin und die Märkischen Wasserstraßen*, Hamburg, 1994.

7   Rohde, H., 'Überlegungen zur mittelalterlichen Wasserstraßen Eider/Treene/Schlei', *Offa*, Vol. 43, 1986.

8   *Archaeological Survey of Egypt* (El Berseh) Part 1, London, 1893.

9   'Vom Bau des Canals von Korinth', *Zbl*, 1891.

11/12 Haasler, W., 'Entwicklung des Wehr- und Schleusenbaues in China', *Bauingenieur*, 1939.

14   After a print by Adriaen van Stalbemt (1588-1622).

15   Historical Topographic Atlas, Gemeentearchief Amsterdam.

20   Leupold, J., *Theatrum Machinarum Hydrotechnicarum*, Leipzig, 1724.

21   Sturm, L. C., Abhandlung von Scläussen und Roll-Brücken, Augspurg, 1715.

22   Paget-Tomlinson, Edward.

24   Dehnert, H., *Schleusen und Schiffshebewerke*, 1954.

25   de la Garde, Jacques.

29   Science Museum Library.

30   Based on Hassall & Trickett, 'The Duke of Bridgewater's Under-ground Canals', *The Mining Engineer*, Oct 1963.

33   Shropshire Records and Research Unit.

34   Dutens, J., *Mémoires sur les Travaux Publics de l'Angleterre*, Paris, 1819 (Science Museum Libr'y).

36   Plymley, *A General View of the Agriculture of Shropshire*, 1803.

39/40 Shropshire Records and Research Unit.

42   Geheimes Staatsarchiv, Berlin.

44   Zabrze Mining Museum Archive, Poland.

45/46 Fulton, R., *A Treatise on the Improvement of Canal Navigation*.

49   National Waterways Museum, Gloucester.

55/56 *Engineering*, 1868.

57   Canal Society of New Jersey.

61   Schmid, 'Der Elbing-Oberländische Kanal', *ZfBau*, 1861.

67   Möller, M., *Grundriss des Wasserbaues*, Leipzig, 1906.

68   Schmid, 'Der Elbing-Oberländische Kanal', *ZfBau*, 1861.

73   Möller, M., *Grundriss des Wasserbaues*, Leipzig, 1906.

75/76 *ZVDI*, 1893.

77   *Engineering*, 1893.

78   The Boat Museum, Ellesmere Port.

79   *Engineering*, 1893.

81/83 Canadian Heritage Parks.

87   de Salis, H. R., *A Handbook of Inland Navigation*, 1901.

90   Schönfelder & Mohr, 'Die Dodge-Schleuse am Cheasepeak-Ohio-Canal', *ZfBau*, 1879.

91   Gerhardt, 'Der Wettbewerb für ein Schiffshebewerk im Donau-Oder-Kanal', *ZBl*, 1905.

93/94 Bodnev, M., in *Recnoj Transport*, 1967.

95   in *Le Génie Civil*, 1964

109 Descombes, R., *Le Plan Incliné de St. Louis/ Arzviller*.

111 Aubert, J., 'Der Ersatz von Schleusen durch das Wasserkeil-Abstiegsbauwerk', *ZfB*, 1973.

112 Greve, J., 'Die geneigte Schleuse', *ZBl*, 1885.

122/124 Wagenbreth, O., 'Der Churprinzer Bergwerks-kanal', *Mitt. des Canal-Vereins*, 16/17, 1996.

125 Model in Stadt- und Bergbaumuseum, Freiberg.

129/130 Dangchiau Water Control Project brochure, 1974.

131 Allsop, N., *The Somersetshire Coal Canal*, 1993.

132 Archive of the Fürstenwalde Waterway Office.

133 Rothmund, L., 'Die Schleuse ohne Wasserverbrauch', *BT*, 1951.

135 *ZfBau*, 1901.

136 *ZfBau*, 1895.

137 *ZfBau*, 1901.

139 Archive of the Westfalian Industry Museum.

141/143 Reinhardt, W., 'Die Hebewerke Rothensee und Hohenwarthe', *BT*, 1938.

144 Archive, Waterway Office Magdeburg.

146 Reinhardt, W., 'Die Hebewerke Rothensee und Hohenwarthe', *BT*, 1938.

147 Archive, Waterway Office Magdeburg.

148 Stütz.

152/154 Wasser- und Schiffahrtsdirection West.

156 Engels, H., *Handbuch des Wasserbaues*, 1923.

159 Bellingrath, E., *Studien über Bau und Betriebs-weise eines deutschen Kanalnetzes*, 1879.

160 'The Anderton Boat Lift', *The Engineer*, 1908.

161 Ernst, A., 'Die Schiffshebewerke bei Les Fontinettes und La Louvière', *ZVDI*, 1890.

165 Ernst, A., 'Hydraulische Hebewerke mit hydrostatischer Ausbalancirung der todten Lasten', *ZVDI*, 1883.

171 Sperling, W., 'Über den Betrieb der Schiffshebewerke des Centrekanals in Belgien', *BT*, 1925.

173 Michael Clarke, Belgium.

174 *ZVDI*, 1907.

176 Canadian Heritage Parks.

178 Archive, Institute of Civil Engineers, London.

181 Green, J., 'Description of the Vertical Lifts … on the Grand Western Canal', *Trans. Inst. Civil Engs.*, 1838.

182 Paget-Tomlinson, Edward.

183 Arnold, Harry.

186 Plarre, 'Vorarbeiten für das Schiffshebewerk Niederfinow', *ZBl*, 1930.

187 Ellerbeck, 'Entwurfsarbeiten für das Schiffshebewerk bei Niederfinow', *Das Schiffshebewerk Niederfinow*, 1935.

190 Ellerbeck, 'Zur Betriebseröffnung des Schiffshebewerk Niederfinow', *Das Schiffshebewerk Niederfinow*, 1935.

191 Ellerbeck, 'Entwurfsarbeiten für das Schiffshebewerk bei Niederfinow', *Das Schiffshebewerk Niederfinow*, 1935.

192 Plarre, 'Vorarbeiten für das Schiffshebewerk Niederfinow', *ZBl*, 1930.

195 Ellerbeck, 'Zur Betriebseröffnung des Schiffshebewerk Niederfinow', *Das Schiffshebewerk Niederfinow*, 1935.

196/197 Archive, Waterway Office, Eberswalde.

199   Bundesministerium für Verkehr.

200 Waterway Office, Uelzen.

203 Infromation from Ministerie van openbare Werken.

2090/210 *ZBl*, 1930.

211/212

213/215/216 *ZBl*, 1930.

Abbreviations for Periodicals & Journals

BT - *Die Bautechnik*

ZfB - *Zeitschrift für Binnenschiffahrt*

WaWi - *Die Wasserwirtschaft*

ZBl - *Zentralblatt der Bauverwaltung*

ZVDI - *Zeitschrift des Vereins Deutscher Ingenieure*

ZfBau - *Zeitschrift für Bauwesen*

Footnotes

1   Rohde, H., 'Von der Treene zur Schlei', in *Flüsse und Kanäle. Die Geschichte der Deutschen Wasserstraßen*, ed. M. Eckoldt. Hamburg 1997.

2   Neuburger, A., *Die Technik des Altertums*, Leipzig 1929, p.207.

3   Hadfield, Charles, *World Canals*, ISBN 0-7153-8555-0, 1986, p.16.

4   Vercoutter, J., 'Excavations at Mirgissa II', in *Kush, Journal of the Sudan Antiquities Service*, Vol XIII (1965), p.68; also Tew, D., *Canal Inclines and Lifts*, ISBN 0-86299-344-X, 1984, p.1.

5   Hadfield (see 3), p.17.

6   Zweig, St., *Sternstunden der Menschheit*, Frankfurt am Main, 1971, p.38, from where the following somewhat over-dramatised description of carrying boats across the two mile wide peninsula is taken: There Mahomet (Sultan and commander of the Turks at the siege of Byzantium) had the ingenious plan to transport his fleet from the outer sea, where it was useless, over the headland to the harbour of the Golden Horn. This breathtaking and bold idea, to carry his ships over a mountainous headland, appears at first absurd and so impossible that the Byzantines and the Genoese of Galata put it as far from their strategy as earlier the Romans and later the Austrians had for the rapid transit of the Alps by Hannibal and Napoleon. Experience says ships can only move in water, and a fleet can never cross a mountain. However, it is the true mark of a demonic will that he transforms the impossible into reality, and always recognizes a military genius, in that it jeers at the rules of war and in the given moment uses creative improvisation instead of tested methods. This is one of the greatest in the annals of the history. Expeditiously Mahomet had countless round pieces of wood brought and made by people into sleds, on which the ships were pulled from the sea as if on a movable dry dock. Simultaneously, thousands undertook excavation work to smooth as much as possible the narrow path over the hill of Pera. To disguise the sudden congregation of so many work people from the enemy, the sultan ordered every day and night a terrible cannon onslaught from Mörsern on the neutral city of Galata, which was ostensibly futile but was actually to divert attention in order to conceal the journey of the ships over the mountains from one sea to the other. While the enemies were busy and only expected attack from the country, ships were drawn over the mountains on the rollers, amply smeared with oil and fat, each on its sled drawn by many buffalo and pushed from behind by sailors from the ships. No sooner had the night arrived, than the wonderful excursion began. Silently, in a clever and premeditated way, the miracle was executed: a whole fleet was carried over the mountain. Decisive in all great military exercises is always the element of surprise. And here in great store was Mahomet's particular genius. Nobody suspected anything of his project ."If a hair in my beard had known my thoughts, I would have torn it out", said this brilliant person, and in most perfect order, while the guns thunder at the walls, his orders were executed. Seventy ships were carried in this one night, the 22nd April, from the sea on one side to the other, over mountain and valley, through vineyards and fields and forests. The next morning the citizens of Byzantium thought they were dreaming: a hostile fleet, as if brought by a ghostly hand, was sailing fully manned with flags flying in the heart of their supposedly unapproachable bay.

7   Feldhaus, F. M., *Die Technik der Vorzeit, der geschichtlichen Zeit und der Naturvölker*, München 1965, p.944 Schneider, Helmuth, 'Die Gaben des Prometheus. Technik im antiken Mittelmeerraum zwischen 750 v. Chr. und 500 n. Chr.', *Propyläen Technikgeschichte*, Berlin 1991. p.155.

8   Hadfield (see 3), p.17.

9   Haasler, W., 'Entwicklung des Wehr- und Schleusenbaues in China', *Bauingenieur* 1939, p.595f.

10   Hadfield (see 3), p.22.

11   Tew (see 4), p.2.

12   *The Book of Ser Marco Polo the Venetian*, ed. H. Cordier, trans. Sir H. Yule, (1903) Vol.ii p.175 n.2.

13   Trout, W. E., 'The Emperor's lock model', *The Best from American Canals*, No. II, ISBN 0-933788-45-2, 1984, p.84.

14   Barrow, J., *Travels in China*, (1804), p.152.

15   Tew (see 4), p.2.

16   Gerdau, B., 'Schiffshebewerke', *ZVDI*, 1896, p.57.

17   Kingma, J., 'Overtoomen in Nederland', *Industriele Archeologie*, No. 39 (1991), p.48-64.

18   Bicker Caarten, A., *Middeleeuwse watermolens in Hollands polderland*, Wormerveer 1990, p.37-39.

19   Ottevanger, G. e.a., *Molens gemalen en andere waterstaat kundige elementen in midden-Delflands*, Den Haag 1985, p.264.

20   *Schuitje varen, theetje drinken varen we naar de Overtoom, drinker er zoete melk met room zoete melk met brokken; kindje mag niet jokken!*

Schute fahren, Tee trinken
fahren wir zum Overtoom,
trinken da süße Milch mit Sahne
süße Milch mit Brocken;
Kindchen darf nicht schwindeln!

21 Kubec, J., Podzimek, J., *Wasserwege*, Hanau 1996, p.15.

22 Leupold, J., *Theatrum Machinarum Hydrotechnicarum*, Leipzig 1724.

23 Sturm, L. C., *Abhandlung/Von Schläussen und Roll-Brücken*, Augsburg 1715.

24 The Friedrich Wilhelms Canal, constructed 1662-1668 between the Spree and Oder, was the first canal link between the river basins of the Elbe and Oder and also the oldest continuous link between the two German river basins.

25 Simons, H., 'Über die Gestaltung von Schiffshebe-werken', *Mitt. der Hannoverschen Versuchsanstalt für Grundbau und Wasserbau*, Franzius-Institut der TH Hannover, Hannover 1957, Vol. 11, p. 5.

26 Partenscky, H.-W., *Binnenverkehrswasserbau-Schiffshebewerke*, Berlin, 1984

27 The following chronological, technical and geographical details about British lifts is taken from the principal works; Tew (see 4) and Paget-Tomlinson, E., *The Illustrated History of Canals & River Navigations*, ISBN 1-85075-276-1, 1993.

28 Hagen, G., *Handbuch der Wasserbaukunst, 2. Theil: Die Ströme*, Vol. 3, Königsberg 1857, p.361.

29 Dutens, J., *Mémoires sur les travaux publics de l'Angleterre*, Paris 1819.

30 Morriss, R. K., *Canals in Shropshire*, 1991, ISBN 0-903802-47-3

31 Hagen, G., *Handbuch der Wasserbaukunst, 2, Theil: Die Ströme*, Vol. 4, Berlin 1874, p.107ff.

32 Dutens, J., (see 29).

33 Brown, I. R., 'Underground Canals in Shropshire Mines', *Mining History*, Vol.13, No.4, Winter 1997, pp17-23

34 Malet, Hugh, *Bridgewater, The Canal Duke 1736-1803*, 1977, ISBN 0 7190 0679 1
Mather, F. C., *After the Canal Duke*, Oxford, 1970

35 Oeynhausen, C. v. and Deschen, H. v., *Schienenwege in England 1826 und 1827*, Berlin 1829, p.167 ff., published in English as *Railways in England 1826 & 1827*.

36 Today they are in the archives of the Mining Museum in Chorzów (Poland).

37 From a personal communication with Dr.-Ing. Martin Eckoldt. The statement comes from the unpublished original version of his dissertation.

38 Hagen (see 31), p.119 ff.

39 Chevalier, *Histoire et description des voies de communication aux Etats-Unis*, Vol. II, p. 476

40 Tew (see 4), p.29.

41 Tew (see 4), p.30.

42 Kalata, B. N., *A Hundred Years, A Hundred Miles*, 1983, ISBN 0-910301-07-7

43 Tew (see 4), p.31.

44 Shank, W. H., *The Amazing Pennsylvania Canals*, 1981, ISBN 0-933788-37-1

45 Schmid, 'Der Elbing-Oberländische Kanal', *ZfBau* 1861, p.149 ff.

46 Hagen (see 31), p.129 ff.

47 Hagen (see 31).

48 Schmid (see 46).

49 'Der Elbing-Oberländische Kanal', *ZfBau* 1885, p.63 ff.

50 Tycner, J., *Der Oberländische Kanal und das Oberland*, Warsaw.

51 *Shubenacadie Canal Guide*, Shubenacadie Canal Commission, 1989.
Information from the Shubenacadie Canal Commission, Dartmouth, Nova Scotia, Canada, *Waterways World*, 1990.

52 *ZVDI* 1893, p.1015-1017.

53 Tew (see 4), p.28.

54 Edwards-May, D., *Binnengewässer Frankreichs*, Hamburg 1993, p.154.

55 'Kioto and Lake Biwa Canal', *Engineering*, 25. Jan. 1889, p.7.
'The Lake Biwa, or Kioto-tu Canal', *Engineering News*, 13th April 1893, p.34.
Trout, W. E., 'A True Account of the Adventures of an American on Japan's Biwako Canal in the Year of the Monkey', *The Towpath Post*, Journal of the Canal Society of New Jersey, Vol.4, Nr.4 1974, p.2ff.

56 Trout (see 55).

57 Crabtree, H., *Railway on the Water*, 1993, ISBN 0 9522592 0 6

58 Cole, J. M., *The Peterborough Hydraulic Lift Lock*, Friends of the Trent-Severn Waterway, 1987, p.5 ff.
Squires, R., 'The Trent-Severn Waterway', *The Best from American Canals*, No. II, ISBN 0-933788-45-2, 1984, p.74.

59 Cole (see 58), p.8.

60 Tew (see 4), p.40.

61 Kubec, J. und Podzimek, J. (see 21), p.304.

62 Partenscky, H. W., (see 26)

63 Brennecke, 'Die Schiffsschleusen', *Handbuch der Ingenieurwissenschaften*, Part III., Vol.8, Leipzig 1904.
Tew (see 4), p.67, …had Solages and Bossu already suggested a plan to build a lift by 1801, in which a water-filled caisson was raised and lowered by means of flotation tanks.

64 Paget-Tomlinson, E., 'West Country Waterway', *Waterways World*, 1984, p.38 ff.

65 Tew (see 4), p.15 f.

66 Leslie, J., 'Description of an Inclined Plane…', ***Proc. Inst. Civil Engs.***, Vol.145, 1853/4, pp205-221

67 Tew (see 4), p.36.

68 Page, D., 'Doon the Gazoon', ***Waterways World***, May 1989, p.72 ff.

69 Page (see 68), p.74

70 Schönfelder und Mohr, 'Die Dodge-Schleuse am Cheasepeake-Ohio-Canal', ***ZfBau***, 1879, p.49 ff

71 Tew (see 4), p.27.

72 Skramstad, H, 'The Georgetown Canal Incline', ***Technology and Culture***, the International Quarterly of the Society for the History of Technology, Chicago, October 1969, Volume 10, No.4, p.549 ff.

73 Gerhardt, P., 'Der Wettbewerb für ein Schiffshebewerk bei Prerau im Donau-Oder-Kanal', ***ZBl***, 1905, p.125 f.

74 Konz, O., ***Neckar-Donau-Kanal, Plochingen-Ulm***, Stuttgart 1954.
Donau-Bodensee-Kanal, Ulm-Friedrichshafen, Stuttgart 1950.

75 Bodnev, M., 'Ein neues Schiffshebewerk am Jenissej', ***Schiff und Hafen***, 1967, H. 10, p.732 ff.

76 Bodnev (see 75)
Partenscky, H.-W., (see 26), p.149 ff.

77 Gallez, A., ***Die schiefe Ebene von Ronquières***, Charleroi, 1972.

78 Simons (see 25), p. 49.

79 Smith, P. L., ***Discovering Canals in Britain***, Buckinghamshire, 1993, p. 57 as well as information published by the Foxton Inclined Plane Trust.

80 Bretschneider, 'Das quergeneigte Schiffshebewerk bei Arzviller', ***WaWi***, 1969.
'The Transverse Incline of St-Louis-Arzviller', ***Review de la Navigation Fluviale Européenne Ports et Industries***, 25 June 1970.

81 Partenscky (see 26), p. 177 ff.

82 Greve, J., 'Die geneigte Schleuse', ***ZBl***, 1885, p.198f.

83 Aubert, J., 'Der Ersatz von Schleusen durch das Wasser-keil-Abstiegsbauwerk', ***ZfB***, 1973, Vol.11, p.511 ff.
Scheuch: 'Die Wasserrutsche von Montech/Garonne', ***Tiefbau***, 1974.

84 Dietrichs, E., 'Das älteste deutsche Schiffshebewerk bei Halsbrücke und der Kanal von Großschirma nach Halsbrüche', ***Wiss. Zs. der Hochschule für Verkehrswesen***, Dresden, 1966, p. 417-421.

85 The difference in level, in which recent literature (Dietrichs, see 84, and Wagenbreth, O., 'Der Churprinzer Bergwerkskanal, das Schiffshebewerk Rothenfurth und weitere Schiffahrtskanäle im Bergbau von Freiberg/Saxony', ***Mitt. des Canalverein***, No. 16/ 17. Rendsburg 1996, p. 15-74.) suggest that it was slightly lower at 7 m. The question is whether Hagen, later to be the greatest German water civil engineer of the 19th Century, was deceived.

86 Hagen (see 28), p. 352.

87 Wagenbreth (see 85), p 34 &. 59.

88 Wagenbreth (see 85), p 32.

89 Wagenbreth (see 85), p 60.

90 Dietrichs see in note 78.

91 Wagenbreth (see 85), p. 43.

92 Crabtree, H. (see 57)

93 From, ***Bescherming waterstaatkundige monumenten in Noord-Holland***, Provinciaal Bestuur van Noord-Holland, Haarlem, 1989.

94 Arends, G. J., ***Sluizen en stuwen***, Delft 1994, p.53.

95 Partenscky (see 26), p. 197 ff.
Information from Schröder, D. & Zimmermann C., 'Wasserstraßen und Binnenschiffahrt in China', ***ZfB***, 1987, Vol.3, p.8 The plant at the Danchiangkou dam has a maximum rise of 86 m and is composed of 5 sections: Lower entrance bay, inclined plane, intermediate canal, vertical lift and upper entrance bay. The maximum rise of the incline is 41 metres, and the vertical lift 45 m. Both dry and wet transport is possible. The caisson for wet is 24 x 10.7 x 1.3 m. For dry movement, boats up to 40 m length can be transported , though they must have a flat bottom. A passage lasts 45 min, and the maximum capacity was determined at 700,000 t/a. Medium-term it is anticipated to increase the capacity of the plant to 300 ton boats.

96 Hagen (see 31), p. 94 f.

97 Tew (see 4), p.62 ff.
Allsop, N., ***The Somersetshire Coal Canal Rediscovered***, ISBN 0-948975-35-0, 1993, p. 90 ff. (which includes the most important newspaper quotations from the building and operating of the caisson-lock).

98 From the Archives of the WSA Berlin, branch office Fürstenwalde.

99 Rothmund, L., 'Die Schleuse ohne Wasserverbrauch', ***BT***, 1951, p. 136 ff.

100 Partenscky (see 26), p. 197.

101 Hagen (see 31), p. 343 f.

102 Tew (see 4) p. 65, Patent no. 1981. to Edward Rowland and Exuperius Pickering

103 Tew (see 4) p.65ff, The lift seems to have been built with a rise of about 3.6 m and could carry 20 ton boats. Its exact location is uncertain.

104 Tew (see 4), p. 67.

105 Dutens (see 29), p. 37.

106 *The Engineer and Machinist*, November 1850, p.259.

107 Faulkner, A., 'A Lock before its Time', *Journal of the Rly & Canal Hist. Soc.*, Vol.31, Pt.8 No.160, March 1995, pp404-410

108 Brennecke, 'Die Schiffsschleusen', *Handbuch der Ingenieurwissenschaft*, III, Part, 8, Leipzig 1904.

109 *Le Genie Civil*, 1892, p. 117.

110 Dehnert, H., *Schleusen und Schiffshebewerke*, Berlin, Göttingen, Heidelberg, 1954, p. 278.

111 'Der Bau des Dortmund-Ems-Canals', *ZfBau*, 1901, p. 278 ff.; *Das Schiffshebewerk bei Henrichenburg am Dortmund-Ems-Kanal*, 1901, reprint available in the museum.

112 de Boer, R., 'Elbquerung über Umweg - Das Wasserstraßenkreuz Magdeburg', *ZfB*, 1995, no. 9, p. 22 ff.

113 These statements were taken from: Reinhardt, W., 'Die Hebewerke Rothensee und Hohenwarthe', *BT*, 1938, p. 618 ff; also *Schiffshebewerk Rothensee*, various guides by the WSA Magdeburg.

114 Illiger, J., 'Das Schiffshebewerk Henrichenburg in Waltrop', *Hansa*, 1963, no. 9, p. 907 ff. as well as Partenscky (see 70), p. 81 ff.

115 Engels, H., *Handbuch des Wasserbaus*, Vol. 2., Leipzig, 1923, p. 1232 f.

116 English patent no. 2498, 1873.

117 Edwards, L.A., *Inland Waterways of Great Britain*, Huntingdon, 1985, p. 372 ff.

118 Sidengham Duer, *Min. of Proc. Inst. Civil Engineers*, XLV (1875-6) pp. 107 et. seq. Ernst, A., 'Hydraulische Hebewerke mit hydrostatischer Ausbalancirung der todten Lasten', *ZVDI*, 1883, p. 329 ff.

119 Carden, David, *The Anderton Boat Lift*, 2000, ISBN 0 9533028 6 5.

120 Hensch, 'Anlagen zur senkrechten Schiffshebung in Frankreich', *ZBl*, 1882, p. 395 ff.

121 Saner, J. A., 'Reconstruction of the Canal-Boat elevator on the River Weaver at Anderton', *Min. of Proc. Inst. Civil Engineers*, XXX, (1909-10), p.239 et seq.

122 Riedler; A.: *Neuere Schiffshebewerke*, Berlin 1897, p. 9.

123 Ernst, A.,' Die Schiffshebewerke bei Les Fontinettes und La Louvière', *ZVDI*, 1890, p.280 ff.

124 Ernst (see 123) and Hensch (see 120).

125 Ernst (see 118).

126 Riedler (see 122), p. 12 f.

127 McKnight, H., *Frankreichs Flüsse und Kanäle*, Bad Soden, 1989, p. 26.

128 Ernst (see 123), p. 284.

129 Sperling, W., 'Über den Betrieb der Schiffshebewerke des Centrekanals in Belgien', *BT*, 1925, p. 501 ff.

130 Ernst (see 123), p. 285.

131 Sperling (see 129), p. 501.

132 Sperling (see 129), p. 505.

133 Dehnert (see 110), p. 275 f. and Cole (see 58).

134 Communication from Mr. John Good

135 The Anderton Boat Lift. Informational material of the Anderton Boat Lift Development Group.

136 Paget-Tomlinson (see 27), p.48.

137 Tew (see 4), p.70.

138 Tew (see 4), p.68 f. and Paget-Tomlinson (see 27), p.48 f.

139 Paget-Tomlinson (see 27), p.258.

140 Tew (see 4), p.72.

141 Green, J., 'Description of the Perpendicular Lifts for passing Boats from one Level of Canal to another, as erected on the Grand Western Canal', *Trans. Inst. Civil Engineers*, 1838, II. p.185 f.

142 Green (see 141), p.188.

143 Hagen (see 28), p.100 f.

144 Paget-Tomlinson (see 27), p.141.

145 Paget-Tomlinson (see 27).

146 Saner (see 121), Weaver Navigation Trustees, 'The Anderton boat lift', *The Engineer*, July 24th, 1908, p. 82 et seq.

147 Saner (see 121).

148 Uhlemann, H.-J., *Berlin und die Märkischen Wasserstraßen*, Hamburg 1994, p.24 ff.

149 *Das Schiffshebewerk Niederfinow*. Special publication of the *BT* and its companion *Der Stahlbau*, years 1927 to 1935, Berlin 1935.

150 *Das Schiffshebewerk Niederfinow*, Published by the WSA Eberswalde 1996.

151 Aster, D., Flaspöler, A. & Schleder, H.-P., 'Das neue Schiffshebewerk Niederfinow', *ZfB*, 1997, Nr. 6, p.31.

152 *Schiffshebewerk Lüneburg in Scharnebeck*, Wasser- und Schiffahrtsdirektion Hamburg 1975, also Partenscky (see 70), p.33 ff.

153 Letter from the des Amtsvorstandes des WSA Uelzen, Herrn BDir Trapp vom 20. März 1990.

154 Corinth, E., 'Schiffshebewerke am Canal du Centre in Belgien', *ZfB* 1987 Nr. 4, p.26 ff., Delmelle, J., *De kabelliften van Strépy-Thieu*, 1986 (Brochure from visitor centre), and Partenscky (see 26), p.58 ff.

155 Schröder, D. u. C. Zimmermann, 'Wasserstraßen und Binnenschiffahrt in China', *ZfB*, 1987, No. 3, p.8.

156 www.bsos.umd.edu/mwr/cwrc/projects.htm www.probeinternational.org/probeint/Three Gorges/tgp/tgp1 9.htm

157 *Das Schiffshebewerk Niederfinow*, (see 150).

158 -ibid-, (see 150), C 6.

159 -ibid-, (see 150), C 6.

160 -ibid-, (see 150), C 9.

Index

A Chronological List of Boat Lifts and Inclines, with Rise, Inclination, Boat Size and Other Information.

(Incorporating some which are not mentioned in the text)

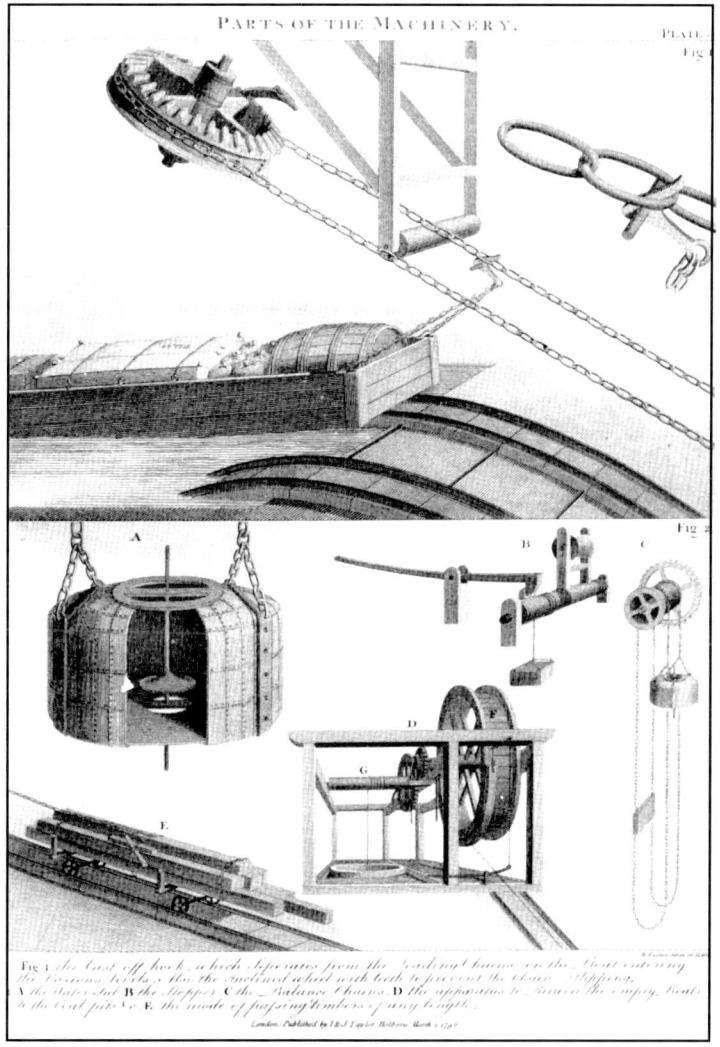

Illustration from Fulton's book, as mentioned on page 37.

Date in operation	Type	Location	Rise	Slope	Boat size	wet or dry	Other information, including builder or designer if known.
England, Frome River							
1760s	Lift	At weirs on river	various	—	boats 10 tons	Crane for containers	Cranes used to pass containers from one boat to another past weirs in the river.
Ireland, Ducart's Canal and Lagan Navigation							
1767	Lift	Coalisland	148 feet	—	used by containers	Crane	Davis Ducart. Original plan was for a vertical shaft for lowering boxes of coal between two canal levels. It was never built.
1777 - 1787		Brackaville	55 feet	?	2 tons, 4.5 ft wide, 10 ft long and 2.5 ft deep	Dry in cradles	Davis Ducart. Double track counter balanced by ascending boat, coal traffic was downhill, horse gin to help boats over reverse slope at top of incline.
1777 - 1787	Incline	Drumreagh	65 feet	?			
1777 - 1787		Fernlough	70 feet	?			
England, St. Columb Canal							
c1773-1781	Incline	Morgan Porth and St. Columb Porth	?	?	Used by containers	Inclines for boxes	John Edyvean. Gravity and horse powered winding gear. Boxes filled from boats then lowered down incline.
England, Ketley Canal							
1788 - c.1816	Incline	Ketley	73 feet	?	Tub boats 8 tons	Dry in cradles	William Reynolds. Double track self-balanced with lock at top for each track. Almost all traffic downhill.
Germany, Kurprinz Mine Canal							
1789-1868	Lift	Halsbrücke	6.8-8 metres	—	8.5 x 1.6 metres 3 tons	Dry	Johann Friedrich Mende. Overhead crane operation.
1789-1868		Rothenfurth	6.8-8 metres				

Date in operation	Type	Location	Rise	Slope	Boat size	wet or dry	Other information, including builder or designer if known.
England, Donnington Wood Canal							
c1765-? (mentioned 1788)	Incline	Hugh's Bridge?	?	?	?	Probably dry in cradles	Possible underground incline between levels as at Worsley. Could have been designed by John Gilbert who probably built Worsley incline. No other details.
1790 - 1873 or 1879	Incline	Hugh's Bridge	42.66 feet	363 feet 1 in 10	Tub boats?	Probably dry in cradles	Originally hoist for boxes (see below), but c1790 replaced by an incline, also for boxes, but which may have been altered later to carry boats of 3 to 5 ton capacity, probably double track, counter balanced, steam engine assisted.
?-c.1794	Lift	Hugh's Bridge	42.66 feet	—	Used by containers	Crane	Tunnel to bottom of vertical shaft. Crane to lift boats to higher level.
England, Shropshire Canal							
1792 - c.1858	Incline	Wrockwardine Wood or Donnington Wood	120 feet	1050 feet	5 tons, 18 ft by 5 ft by 2.5 ft deep	Dry in cradles	William Reynolds. Double track counter balanced, reverse slope at top. horse gin used at first but almost immediately after the plane was opened a steam engine was substituted.
1792 - c.1858		Windmill Farm	126 feet	1800 feet			
1792 - c.1894		The Hay, Coalport	207 feet	1050 feet			
1794–c.1800	Incline	Brierly Hill, Coalbrookdale	120 feet	?	Used by containers	Incline for boxes	Replaced counter balanced vertical shaft which had operated from 1792. Incline double track counter balanced.
England, Shrewsbury Canal							
1793 - 1921	Incline	Trench	75 feet	669 feet	5 tons	Dry in cradles	William Reynolds. Double track counter balanced, reverse slope at top. steam engine assisted, last to work in Britain
U.S.A., South Hadley Canal, Connecticut River							
1790s-1805	Incline	South Hadley	53 feet	230 feet at 1 in 4	?	Dry in caisson	Benjamin Prescott. Water wheel powered, single track
England, Shropshire Canal							
1794	Lift	Oakengates	1.5 feet		?	Experimental	Robert Weldon. Boat carried in enclosed caisson which floated in water to support main weight.
Wales, Ellesmere Canal							
1796-1799 or 1800	Lift	Probably near Ruabon	12 feet	—	28 tons, narrow boats	Experimental	Edward Rowland and Exuperius Pickering. Boat carried in water filled caisson supported by a float in a column of water. (as at Henrichenburg), rack and pinion operated

Date in operation	Type	Location	Rise	Slope	Boat size	wet or dry	Other information, including builder or designer if known.
England, Bridgewater Canal							
1797-1822	Incline	Ashton's Field, Worsley underground canal system	106.5 feet	453 feet 1 in 4	various up to 12 tons	Dry in cradles	John Gilbert? Double track divided by brick wall with safety openings, counter balanced, self acting, but manually assisted initially by winches and descent controlled by brake. 2 locks at top 54 feet in length. Carriages 30 feet by 7.33 feet.
England, Somersetshire Coal Canal							
1797-1799	Lift	Combe Hay	46 feet	—	Caisson, 80 ft by 10.5 ft by 11.5 ft	Experimental	Robert Weldon. Caisson lock, narrow boat in enclosed caisson in water filled lock chamber which gives buoyancy, rack and pinion operated. Tunnel to lower entrance.
England, Dorset and Somerset Canal, Nettlebridge Branch							
1800-1802	Lift	Barrow Hill	21 feet	—	10 ton boats	Experimental	James Fussell. Balanced caissons into which boats floated. Manually worked but dependant upon water added to descending caisson. 4 more lifts proposed.
Poland (Prussia), Klodnice Canal							
1806-24	Incline	Royal Ironworks Gliwice	5 metres	?	6.3 x 1.37 metres 4 tons	Dry in cradles	Development of the incline in the mines at Worsley with lock at summit.
1806-39	Incline	Sosnice	11.6 metres	?			
France, Canal du Creusot							
1806	Incline-/Lift	3 inclines & 3 lifts	Inclines 8.61, 5.63 & 7.31 metres	Lifts 8.45, 5.36 & 4.55 metres	8 tons	Dry in cradles (inclines)	M. E. M. Gauthey. Water wheel powered, design similar to Fulton's. Lifts designed by M. de Solages. One lift and one incline built. Inclines with wet caisson proposed to replace lifts, but canal not completed on Gauthey's death. Lifts similar to Ruabon lift.
England, Worcester and Birmingham Canal							
1808-1815?	Lift	Tardebigge	12 feet	—	caisson 72 x 8 x 3.5 feet, 20 tons	Experimental	John Woodhouse. Boat floated in a caisson, balanced by brick counterweights, suspended by chains over wheels and moved by manual windlass.
England, Regent's Canal							
1815-1816	Lift	Camden Town	6.66 feet	—	Barges, probably 80 x 14.5 feet	Experimental	Colonel Sir William Congreve. Counter balancing inverted caissons dependant for movement on air pressure and hydraulic rams,

Date in operation	Type	Location	Rise	Slope	Boat size	wet or dry	Other information, including builder or designer if known.
Wales, Redding Canal							
1818	Incline	Nr. Cadoxton	?	?	?	?	Marked on map. Not known if used by boats.
England, Tavistock Canal							
1819-between 1831 & 1844?	Incline	Mill Hill	19.5 feet	936 feet	4.5 ton tub boats 30 feet by 4.5 feet by 2.5 feet	Dry in cradles	John Taylor. Horse hauled single track.
1817-1883	Incline	Morwellham Quay	240.5 feet	720 feet	—	Not for boats	John Taylor. Incline carrying trucks down from canal to riverside quay. Waterwheel assisted.
England, Bude Canal							
1819-1891	Incline	Marhamchurch	120 feet	836 feet			First proposed in 1774 by John Edyvean, then by Edmund Leach in 1785. Finally built by James Green. Waterwheel driven
1819-1891	Incline	Hobbacott Down	225 feet	?			James Green. 2 Buckets, 10 feet dia by 5.5 feet, holding 15 tons, operating in well. Small steam engine also provided.
1819-1891	Incline	Vealand (Venn)	58 feet	500 feet	5 ton boats with wheels, 20 feet by 5.5 feet	Endless chain double track	
1819-1891	Incline	Merrifield	60 feet	360 feet			
1819-1891	Incline	Tamerton	59 feet	360 feet			James Green. Waterwheel driven for wheeled boats
1819-1891	Incline	Werrington or Bridgetown	51 feet	259 feet			
England, Torrington or Rolle Canal							
1827-1871	Incline	Weare Giffard	?	?	Probably wheels on the boat	Endless chain double track	James Green. Waterwheel driven, probably for wheeled boats.
U.S.A., Morris Canal							
1828-1924	Incline	23 inclines	100 to 35 feet	150 to 50 feet	Orig. 25 tons	Dry on cradle	James Renwick. Originally waterwheels, later turbines, some with locks, others sill at higher end. In 1835-6 all locks, reverted to sills, work undertaken in 1845-60. Three double inclines, the rest single
1903-1924	Incline	Orange Street, Newark	approx 7 feet	?		Dry on cradle	Electrically powered incline over railway which passed underneath the canal at this point.

Date in operation	Type	Location	Rise	Slope	Boat size	wet or dry	Other information, including builder or designer if known.
England, Grand Western Canal							
1836-1837	Incline	Wellisford	81 feet	440 feet 1 in 5.5		Floating in caisson	James Green. 2 by 10 ton buckets in well operation, caissons connected to endless chains. Steam engine installed 1838.
1834-1867	Lift	Taunton	23.5 feet	—	8 ton tub boats 26 feet by 6.5 feet by 2.25 feet	Floating in caisson	James Green. Two caissons which counter balanced each other, water added to top caisson and drained when it reached lower position, manual operation if needed.
		Norton Fitzwarren	12.5 feet	—			
		Allerford	19 feet	—			
1835-1867	Lift	Trefusis	38.5 feet	—			
		Nynehead	24 feet	—			
		Winsbeer	18 feet	—			
		Greenham	42 feet	—			
England, Chard Canal							
1842-1868	Incline	Thornfalcon	28 feet	c. 1 in 8		Floating in caisson	Sydney Hall or Sir Wm. Cubbitt? Double track, counter balanced, water added to descending caisson, brake fitted
1842-1868	Incline	Wrantage	27.5 feet	c. 1 in 8	8 ton tub boats, 26 feet by 6.5 feet by 2.25 feet		
1842-1868	Incline	Ilminster	82.5 feet	c. 1 in 9			Sydney Hall or Sir Wm. Cubbitt? Endless chain, powered by water wheel at bottom of incline
1842-1868	Incline	Chard Common	86 feet	c. 1 in 8		Dry on cradle	Sydney Hall or Sir Wm. Cubbitt? Single track with incline, turbine powered
Wales, Kidwelly and Llanelly Canal							
1838-1867	Incline	Pont Henry	57 feet	1 in 6	6 ton boats	Probably dry on cradle	James Green. Probably counter balanced with hydraulic pumps.
1838-1867	Incline	Capel Ifan	56 feet	1 in 6			
1838-1867	Incline	Hirwaun-isaf	83 feet	1 in 10			James Green. Probably counter balanced with hydraulic pumps. never completed

Date in operation	Type	Location	Rise	Slope	Boat size	wet or dry	Other information, including builder or designer if known.
Scotland, Monkland Canal							
1857-1887	Incline	Blackhill	96 feet	1040 feet 1 in 10	Scows up to 70 tons	Floating in caisson	James Leslie. Double track, counter balanced, descending caissons filled with water, steam assisted, vertical lifting gates Later caissons drained due to surging.
Poland (originally Prussia), Elblanski Canal (Elbing Oberland Kanal)							
1860-		Buczyniec (Buchenwald)	20.42 metres				Georg Steenke. Double track, counter-balanced, water wheels as power source.
	Incline	Katy (Kanten)	18.90 metres	1:12, 1:24 on first section	24.5 by 3.0 by 1.1 metres 50 tons	Dry on cradle	
		Olesnica (Schönfeld)	24.39 metres				
		Jelenie (Hirschfeld)	21.95 metres				
1881-		Calony Nowe (Neu-Kussfeld)	13.72 metres				Double track, counter-balanced, turbine as power source
Canada (Nova Scotia), Shubenacadie Canal							
1861-70	Incline	Portobello	35 feet	600 feet 1 in 16	66 feet by 16.5 feet by 4 feet	Dry on cradle	Charles W. Fairbanks. Turbine powered similar to Morris Canal.
1861-70	Incline	Dartmouth	60 feet	1,500 feet 1 in 22			
England, Aire & Calder Navigation							
1863-1986	Lift	Compartment boat hoists at Goole	12 - 20 feet	—	40 ton compartment boats	Dry on cradle	William Bartholomew. Hydraulically powered boat lift. Five built between 1863 and 1912, last one in service until 1986.
England, Weaver Navigation and Trent & Mersey Canal							
1875-1986	Lift	Anderton	50 feet	—	Up to 50 ton craft	Floating in caisson	Edwin Clark. Counter balancing caissons and hydraulic rams, 1906-8 electrified with independent caissons counter balanced.

Date in operation	Type	Location	Rise	Slope	Boat size	wet or dry	Other information, including builder or designer if known.
U.S.A., Chesapeake and Ohio Canal							
1876-1889	Incline	Georgetown	40 feet	600 feet 1 in 12	112 x 16.75 x 7.8 feet, 112 ton boats	Originally floating into caisson	William Rich Hutton. Single track, counter balanced, powered by Leffel water turbine. Later caisson used as cradle for dry operation.
France, Canal de l'Oureq							
1884-1922	Incline	Meaux	12.17 metres	550 metres	boat carries 60 tonnes, total 110 tonnes	Dry on cradle	M. Agudio. Originally chain hauled, then rack system. Turbine powered.
France, Canal de Neuffossé							
1888-1967	Lift	Les Fontinettes	13.13 metres	—	caisson 40 x 5.8 x 2.1 metres boats 300 tons	Floating in caissons	Edwin Clark. Counter balancing caissons and hydraulic rams.
Belgium, Canal du Centre							
1888-	Lift	Houdeng-Goegnies or La Louvié're	15.4 metres	—	400 ton boats	Floating in caissons	Edwin Clark. Twin hydraulic lift based on Anderton and Les Fontinettes
Japan, Biwako Canal							
1894-1914	Incline	Fushimi	?	?	?	Dry on cradle	Sakuro Tanabe. Double incline, counter balanced, water powered. Removed 1968.
1894-1914	Incline	Kyoto	118 feet	1,815 feet 1 in 15	47.5 feet by 7.5 feet by 4 feet	Dry on cradle	Sakuro Tanabe. Double track, counter balanced, electrically powered. Still in place.
Germany, Dortmund-Ems Canal							
1899-1962	Lift	Henrichenburg	14 -16 metres	—	67 x 8.2 x 2 metres 700 tons	Floating in caisson	Jebens. Flotation tank supported single caisson lift.

Date in operation	Type	Location	Rise	Slope	Boat size	wet or dry	Other information, including builder or designer if known.
England, Grand Junction Canal							
1900-1910	Incline	Foxton	75.5 feet	307 feet 1 in 4	Narrow boats & barges up to 50 tons	Floating in caissons	Barnabas J. Thomas & Joseph J. Taylor. Two counter balanced caissons, steam assisted.
Canada, Trent-Severn Waterway							
1904-	Lift	Peterborough	65 feet	—	Boats up to 800 tons	Floating in caissons	Richard B. Rogers. Counter balancing caissons and hydraulic rams,
1907-	Lift	Kirkfield	49 feet				Counter balancing caissons and hydraulic rams.
1919-	Incline	Big Chute	58 feet	?	60 feet by 13.5 feet by 4 feet, 20 tons	Dry on cradle	Single incline, electrically powered. Enlarged incline opened 1978 for 100 ton boats
1919-1965	Incline	Swift Rapids	47 feet	?			Single incline, electrically powered.
The Netherlands, Amsterdam canal system							
1916-1954	Lift	Postjeswetering	2-3 metres	—	14 x 3 x 0.7 metres, 15 tons	Dry on cradle	Overhead crane type.
Belgium, Canal du Centre							
1917-	Lift	Houdeng-Aimeries	16.93 metres	—	400 tons	Floating in caissons	Twin caissons, counter-balanced hydraulic operation.
1917-		Bracquegnies					
1917-		Thieu					
The Netherlands, Zuiderzee link							
1923-	Lift	Broekerhaven	2-3 metres	—	14 x 3 x 0.7 metres, 15 tons	Dry on cradle	Overhead crane type. In operation until 1981, and returned to service after restoration in 1992.
Germany, Havel-Oder Waterway							
1934-	Lift	Niederfinow	36 metres	—	80 x 9 x 2 metres 1,000 tons	Floating in caisson	Single caisson counter-balanced operation.
Germany, Mittelland Canal							
1938-	Lift	Rothensee	18.67 metres max	—	80 x 9 x 2 metres 1,000 tons	Floating in caisson	Flotation lift with single caisson.

Date in operation	Type	Location	Rise	Slope	Boat size	wet or dry	Other information, including builder or designer if known.
Germany, Dortmund-Ems Canal							
1962-	Lift	Henrichenburg	14.5 metres	—	80 x 9.5 x 2.5 metres, 1,350 tons	Floating in caisson	Flotation lift with single caisson
Czech Republic, Moldau River							
1962-	Incline	Orlik Dam	71.5 metres	c1:2	300 ton boats	Floating in caisson	Not completed
1962-	Incline	Orlik Dam	71.5 metres	c1:2	3.5 ton pleasure boats	Dry on cradle	The wagon and cradle are rotated through 180 degrees at the dry summit of the inclines.
England, Aire & Calder Navigation							
1964-	Lift	Ferrybridge	12.2 metres	—	170 compartment boats	Dry on cradle	Unloading system at power station for compartment boat system developed from Tom Puddings.
China, Lushui River							
1967-	Lift	Puji	26 metres	—	16.1 x 3.8 metres 20 tons	Dry on cradle	Overhead crane type lift
Russia, Yenisey River							
1968-	Incline	Krasnoyarsk	100-108 metres	1:10 306 metres and 1,189 metres	caisson 90 x 18 x 3.3 metres, 2,000 tons	Floating in caisson	Caisson rotates on turntable at summit, gate at one end of the caisson only.
Belgium, Canal de Charleroi à Bruxelles							
1968-	Incline	Ronquières	68 metres	1:20	caisson 87 x 12 x 3.7 metres, 1,350 tons	Floating in caisson	Double lift, each caisson with counterweight, electrically powered.
France, Canal de la Marne au Rhin							
1969-	Incline	Arzviller	44.55 metres	1:2.44 136 metres	caisson 41.5 x 5.5 x 3.2 metres 350 tons	Floating in caisson	Designed as double incline, but only one caisson built. Electric drive with caisson counter balanced by weights.
China, Hanjiang River							
1973	Incline	Danjiangkou	33.50 metres	?	platform 32 x 10.7 metres, caisson 24 x 10.7 x 0.9 metres, 150 tons	Both in the dry or in a reduced length caisson	Rise could be 41 metres.
1973	Lift		35 metres	—			Overhead electric crane type lift, rise could be 45 metres.

Date in operation	Type	Location	Rise	Slope	Boat size	wet or dry	Other information, including builder or designer if known.
France, Canal Lateral à la Garonne							
1973-	Incline	Montech	14.3 metres	1:33.3	350 tons	Water-slope	Diesel-electric tractors.
Germany, Elbe-Seiten Canal							
1975-	Lift	Scharnebeck	38 metres	—	80 x 9.5 x 2.5 metres 1,350 tons	Floating in caisson	Double lift, with two caissons with counter-weights.
France, Canal du Midi							
1983-	Incline	Fonserannes	13.6 metres	1:33.3	350 tons	Water-slope	Diesel-electric tractor.
China							
1990-	Lift	Angkang	87.9 metres	—	24 x 7 metres 100 tons	Sitting on platform	Overhead crane type lift
China, Mingjiang River							
1998-	Lift	Shuikou	59 metres	—	114 x 12 x 2.5 metres 2 x 500 tons	Floating in caisson	Counter-balanced lift with one caisson.
China, Hongshui River							
2000-	Lift	Yantan	68.5 metres	—	40 x 10.8 x 1.8 metres 250 tons	Floating in caisson	Counter-balanced lift with one caisson.
Belgium, Canal du Centre							
2001-	Lift	Strepy-Thieu	73.15 metres	—	80 x 9.5 x 2.5 metres 1,350 tons	Floating in caisson	Double lift, with two caissons with counter-weights. Also designed to accommodate push-tow units of 110 x 11.4 x 2.8 metres.
Scotland, Forth & Clyde and Union Canals							
2002-	Lift	Falkirk	20 metres	—	22.5 x 5 metres	Floating in caisson	Circular rotating lift.
China, Qinjiang River							
2002-	Lift	Geheyan	42 metres	—	300 ton	Floating in caisson	Counter-balanced lift with one caisson.